SIMULATION OF ELECTROCHEMICAL PROCESSES

WIT*PRESS*

WIT Press publishes leading books in Science and Technology.
Visit our website for new and current list of titles.
www.witpress.com

WIT*eLibrary*

Home of the Transactions of the Wessex Institute.
Papers presented at Simulation of Electrochemical Processes are archived in the WIT eLibrary in volume 48 of WIT Transactions on Engineering Sciences (ISSN 1743-3533).
The WIT eLibrary provides the international scientific community with immediate and permanent access to individual papers presented at WIT conferences.
Visit the WIT eLibrary at www.witpress.com.

Electrical Engineering and Electromagnetics
Editorial Board

FIRST INTERNATIONAL CONFERENCE ON
SIMULATION OF ELECTROCHEMICAL PROCESSES

ELECTROCOR

CONFERENCE CHAIRMEN

C.A. Brebbia
Wessex Institute of Technology, UK

V.G. DeGiorgi
Naval Research Laboratory, USA

R.A. Adey
Wessex Institute of Technology, UK

INTERNATIONAL SCIENTIFIC ADVISORY COMMITTEE

Sponsored by
Office of Naval Research Global

Organised by
Wessex Institute of Technology, UK

SIMULATION OF ELECTROCHEMICAL PROCESSES

Editors:

C.A. Brebbia
Wessex Institute of Technology, UK

V.G. DeGiorgi
Naval Research Laboratory, USA

R.A. Adey
Wessex Institute of Technology, UK

WIT*PRESS* Southampton, Boston

C.A. Brebbia
Wessex Institute of Technology, UK

V.G. DeGiorgi
Naval Research Laboratory, USA

R.A. Adey
Wessex Institute of Technology, UK

Published by

WIT Press
Ashurst Lodge, Ashurst, Southampton, SO40 7AA, UK
Tel: 44 (0) 238 029 3223; Fax: 44 (0) 238 029 2853
E-Mail: witpress@witpress.com
http://www.witpress.com

For USA, Canada and Mexico

Computational Mechanics Inc
25 Bridge Street, Billerica, MA 01821, USA
Tel: 978 667 5841; Fax: 978 667 7582
E-Mail: infousa@witpress.com
http://www.witpress.com

British Library Cataloguing-in-Publication Data

A Catalogue record for this book is available
from the British Library

ISBN: 1-84564-012-8
ISSN: 1746-4471 (print)
ISSN: 1743-3533 (on-line)

The texts of the papers in this volume were set individually by the authors or under their supervision. Only minor corrections to the text may have been carried out by the publisher.

This work relates to the Department of the Navy Grant N00014-04-1-1031 issued by the Office of Naval Research International Field Office. The United States Government has a royalty-free licence throughout the world in all copyrightable material contained here.

Printed in Great Britain by Athenaeum Press Ltd., Gateshead.

Preface

This book contains most of the papers presented at the first International Conference on the Simulation of Electrochemical Processes held in Cadiz, Spain in May 2005. The meeting was organised by the Wessex Institute of Technology.

The motivation for the meeting was to bring together researchers who have made significant developments in the area of Electrochemical modelling over recent years. Electrochemical processes are used by engineers to protect structures against corrosion, to apply coatings and paints, and as a manufacturing process. However, until recently, engineers had to use experimental testing or frequent surveys to ensure the adequacy of a design as sophisticated prediction models were not available. The papers presented at this conference demonstrate the major advances that have been made in computational modelling to enable the most complex processes to be simulated.

The papers in this book are divided into the following main topics:

- Cathodic Protection Systems
- Modelling Methodologies
- Electrodeposition and Electroforming
- Modelling Coatings
- Modelling Stress Corrosion Cracking and Corrosion Fatigue

The editors are indebted to the members of the International Scientific Advisory Committee for their help in reviewing the abstracts and papers. The generous support of the Office of Naval Research Global is also much appreciated.

The Editors
Cadiz, 2005

Contents

Section 3: Electrodeposition and Electroforming

Section 4: Modelling of coatings

Section 5: Modelling stress corrosion cracking and corrosion fatigue

Section 1
Cathodic protection systems

Computer simulation of the interference between a ship's and a dock's cathodic protection systems

E. Santana-Diaz & R. Adey
Computational Mechanics BEASY, UK

Abstract

Paint and impressed current cathodic protection (ICCP) systems are used to control corrosion on many different types of structures. Since cathodic protection (CP) systems were first applied, engineers have used experience and intensive monitoring to optimise their design to prevent corrosion. Numerical methods and in particular Boundary Elements Methods (BEM) have been widely used in the cathodic protection field to simulate the performance of the CP system and to predict the associated electric and magnetic fields.

Special care must be taken when a new structure is commissioned as additional stray currents could alter the performance of existing CP systems. In this work, the influence of the cathodic protection system of a ship on a dock is predicted using computer modelling. The damage that the cathodic protection system of the ship could cause to the structure is highlighted and possible mitigation measures assessed using the optimisation features of the modelling software.

The data requirements and practical application of computer modelling is discussed and other possible applications highlighted.

1 Introduction

Boundary Element Methods (BEM) have been used to simulate the behaviour of cathodic protection systems since the late 70's [1]. As the name implies, the method requires elements to be created, but only on the boundary (or surfaces) of the problem geometry.

The advantages of boundary elements for CP analysis are many, for instance:

- The meshes are only on the surface, hence only surface elements are required. Mesh generators can be used with confidence, and models can be constructed extremely quickly and inexpensively.

- BE gives the solutions on the boundary and, only if required, at specified internal points. Since for CP analysis the solution is only needed on the surfaces, it is far easier to analyse the results than for FE analysis which automatically gives results for all nodes (internal or boundary).
- BE methods are very effective and accurate for modelling infinite domains as is the case for CP analysis.

2 State of the art

Numerical methods have been widely used in the corrosion field since the early 80's when IMI Marstons employed them to help model the performance of the impressed anodes to be used on the Conoco TLP platform for the North Sea [2].

They have been successfully compared with physical scale modeling experimental results and data obtained from tests performed on full size ships [3]. DeGiorgi, Kee and Thomas studied a 3D model of a U S Navy CG class ship [4] using boundary elements. The potential profiles for reference cell readings of -0.85 Volts Ag/AgCl showed very good agreement between experimental and computational results. The difference between experimental and computational results was 6% for total current values, 5% for Amps to the propeller and 10% for Amps to the docking blocks. The general conclusions were that, *boundary element methods were a viable technique for determining marine corrosion parameters*.

Gartland, Bjoernaas and Osvoll described in their paper published in the 1999 Corrosion Conference in San Antonio, where a special session was organised on computer modelling, what they had learnt in 15 years of work [5]. Some of the points highlighted were:

- *CP design is an area where experience is quite important*, and even if there are several guidelines available, these are not written to meet the new challenges. Computer modelling has no such limitations.
- *Computer modelling is very well suited for "what if-studies"*, where results can be obtained rather quickly compared to time-consuming experiments, or where full-scale experiments cannot even be carried out.
- *The usefulness of CP modelling is perhaps most evident for problems related to uneven anode distribution or hybrid systems.* CP design guidelines offer absolutely no means to create a safe CP design for a structure with uneven anode distribution. All equations and formulas used to calculate the required anode mass and anode size are based on the assumption that the anodes shall be distributed "most evenly" on the structure.
- *The influence of the boundary conditions depend very much on the problem.* For a verification study of a structure covering the entire lifetime, the boundary conditions must be very accurately modelled, which includes their non-linearity and the time dependence.
- Upgrading of CP systems for existing structures is an area where computer modelling has proven to be cost effective. In a study of the Ninian Northen it was concluded that a saving was in the order of

£7.75m using CP Modelling technology combined with measurements of the current densities on the platform before retrofitting.

3 Stray currents

Stray current corrosion can be caused by the systems used to protect ship hulls and buried structures, by the DC electrical systems used to power trolley cars, and by a number of other sources. Stray current corrosion is of immense economic importance, since it has led to such problems as rapid failure of buried natural gas pipelines and water mains, and costly repairs to ship structures and piers.

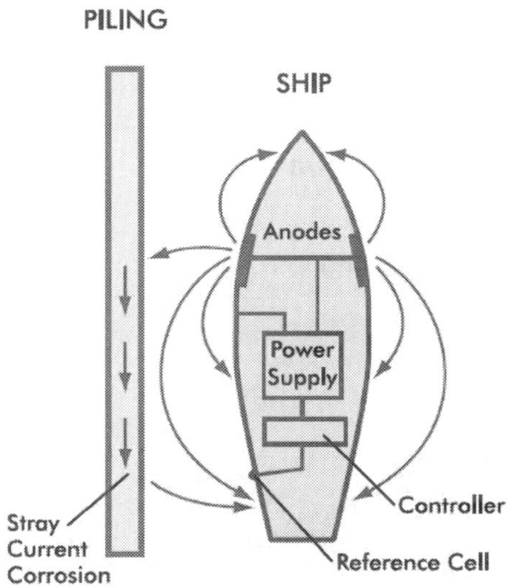

Figure 1: Stray current corrosion of the ICCP system of the ship to the dock.

The basis of the interference model presented in this paper is explained in detail by Trevelyan and Hack [9]. A ships impressed current cathodic protection (ICCP) system consists of a power supply which forces current out of the anode, through the seawater, and into the ship's hull to complete the circuit. The action of the current causes a shift in electrochemical potential of the steel of the ship's hull which slows or stops the normal tendency of steel to corrode in seawater. This technique is used to control corrosion on all large ships and most small ones. On large ships, the current being delivered by the anodes is typically controlled by monitoring the amount of potential shift of the steel hull using special electrodes called reference cells. The current is usually adjusted to give a fixed steel potential near the controlling reference cell. A ship may dock near a large metallic object such as the steel sheet piling on the pier to which the ship is tied. This creates the situation, illustrated in Figure 1. Current flowing from the ship's anode through the seawater to the ship hull encounters a resistance due to the resistivity of seawater. If the sheet piling is closer to the anode than sections of the ship hull, the current has an alternate path that it may follow. It can travel through a much shorter seawater path to the piling, through the low resistance steel of the piling until it gets close to the ship hull, then through another short seawater path between the piling and the hull to complete the circuit. At the point where the current enters the piling, the steel is protected from corrosion the same way as the ship hull is protected by the cathodic protection system. Where the current leaves the piling to enter the seawater, the opposite occurs. The corrosion of the steel piling in this area is dramatically increased. The magnitude of the resulting stray current corrosion depends on how much the cathodic protection current takes the alternate path through the piling.

The behaviour of the cathodic protection system of a frigate in open sea and its influence, interference on a steel made dock, is shown in the next sections of this paper. In addition, a possible mitigation of the interference by placing a sacrificial cathodic protection system on the walls of the dock is also shown.

4 Frigate model, description

The dimensions of the frigate are:

- Waterline length: 34.0m.
- Draft: 2.3m.
- Waterline beam: 6.4m.

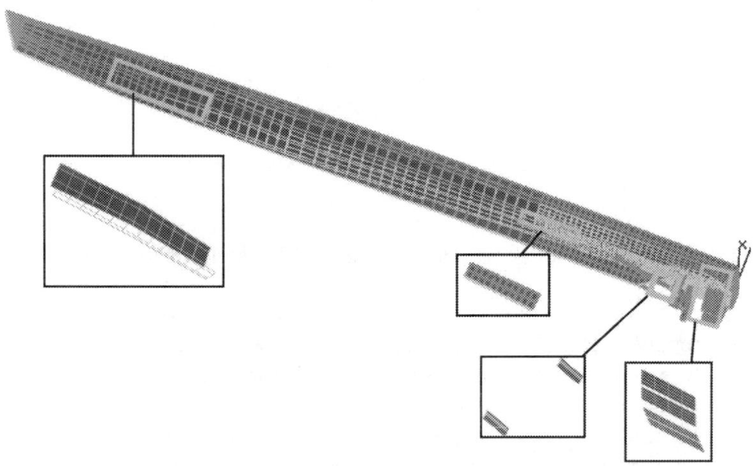

Figure 2: Frigate and damage distribution.

The frigate was considered fully coated apart from some specific damaged areas (Figure 2):

- Area at the stern of the frigate, nearby the propeller, area= 3.51m^2.
- Shaft of the propellers, area= 0.18m^2.
- Small area of the keel, area= 3.18m^2.

- Area at the bow, area= 11.56m^2.

The propeller was set made of Nickel-Aluminium-Bronze with an area of 1.84m^2. The electrolyte considered was seawater with a resistivity of 20ohm·cms [8], what implies a conductive of about 5S/m [8].

Two impressed anodes were placed at the starboard and port sides of the hull of the frigate in addition to two reference cells of silver chloride (Ag/AgCl/seawater). The criterion for protection was that the reference cells should be in -900mV to assure that the surrounded areas were protected.

5 Analysis of the Frigate ICCP system on open sea

Initially the frigate was analysed considering an open sea environment where there were no interference with the environment. Thus, the normal behaviour of the ICCP (Impress current cathodic protection) system could be set.

The current of the impressed anodes were iteratively adjusted to satisfy the potential required at the reference electrodes (-900mV). This was achieved using the boundary elements software assisted by optimisation software [10-14].

The final potential distribution on the hull of the ship is shown below (Figure 3). The area near the propeller is the anode more likely of suffering from corrosion since its potential is above the rest of the areas of the ship.

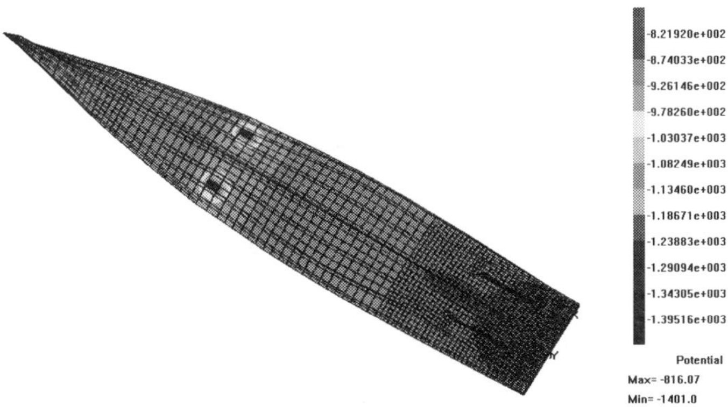

Figure 3: Potential distribution of the frigate in open sea.

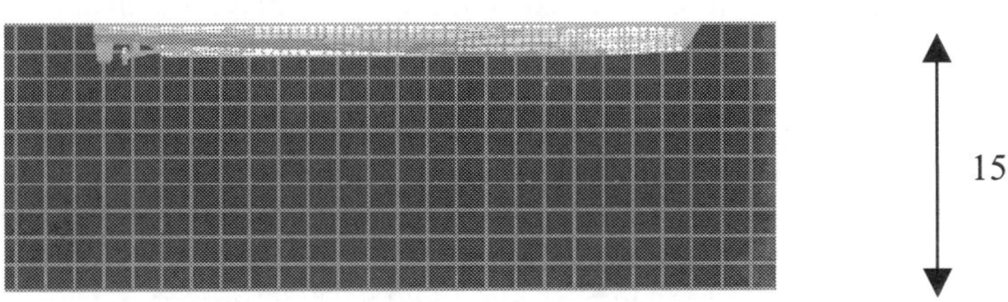

Figure 4: Dock height.

6 Analysis of the Frigate ICCP system berthed at a steel dock

The frigate is moved near a dock where it is berthed for simple maintenance. For the sake of simplicity, the dock is considered a plate made of steel of 15m height.

The frigate was berthed at the port at about 1.75m from the dock. The frigate's ICCP system was not switched off unless a diver needed to inspection the hull.

Figure 5: Distance from the ship to the dock.

The potential distribution on the hull of the ship is shown in Figure 6. It can be observed that the potential distribution presents a slightly higher value than the values had when the ship was on open sea.

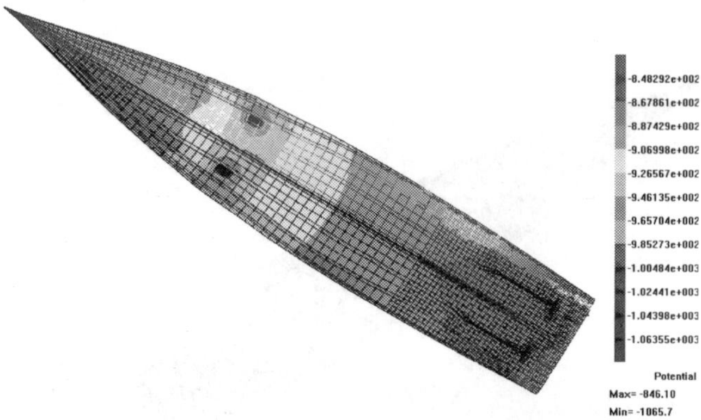

Figure 6: Potential distribution of the frigate's hull berthed at the dock.

The total current supplied by the anodes is about 4% larger than the value had when the frigate was on open sea. Therefore, the dock must be influencing the measurements of the reference cells of the ship to make the ICCP system to generate more current to keep them at -900 mV.

A study of the dock shows that there are two clear areas with different potentials, one of the areas is acting anodically as the other is acting cathodically. In fact, the current the dock is receiving current from the impressed anodes of the frigate and the current returns to the frigate near the propeller, creating two areas with cathodic and anodic behaviour respectively.

The current returning to the ship from the dock will provoke faster corrosion of this area of the dock and therefore a remedy to this situation must be considered.

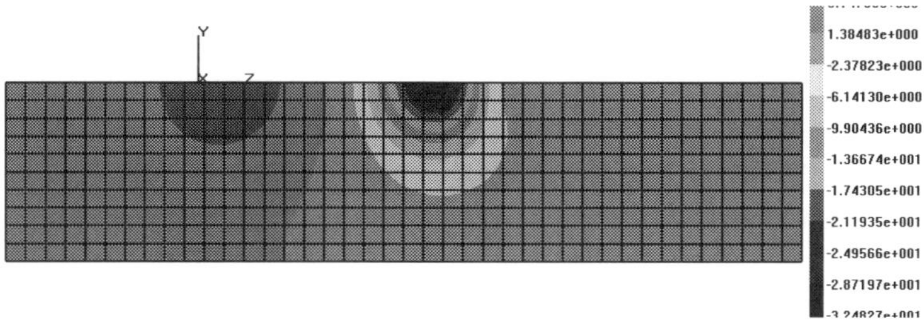

Figure 7: Current distribution on the dock.

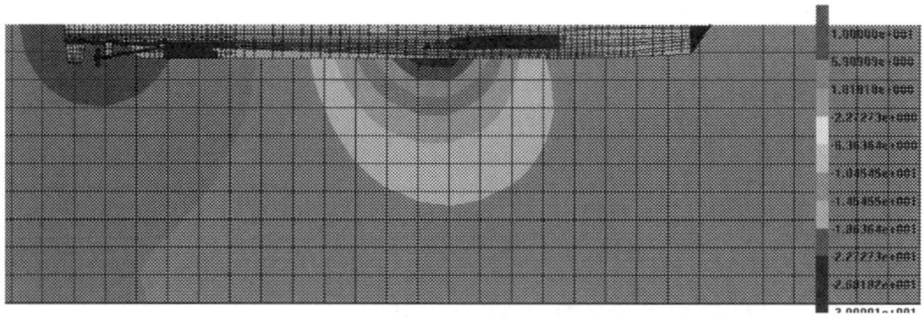

Figure 8: Current distribution on the dock and ship's hull.

7 A passive cathodic protection on the dock

In order to avoid the previous situation in which the dock was being corroded by the ICCP system of the frigate, a passive system (sacrificial anodes) of protection was designed which consisted of an array of 60 zinc anodes was placed on the wall of the dock. The dimensions of those anodes are shown in Figure 9.

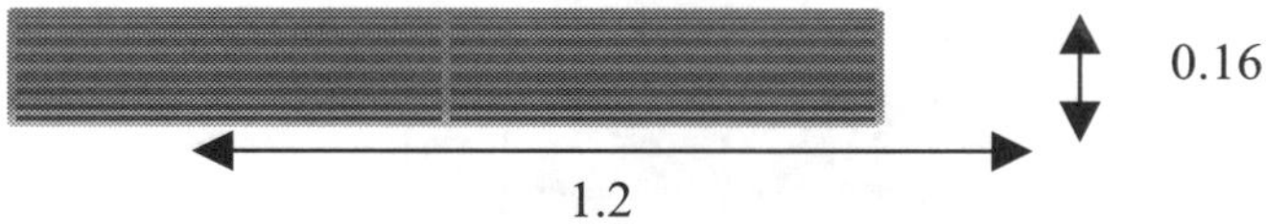

Figure 9: Zinc anode dimensions.

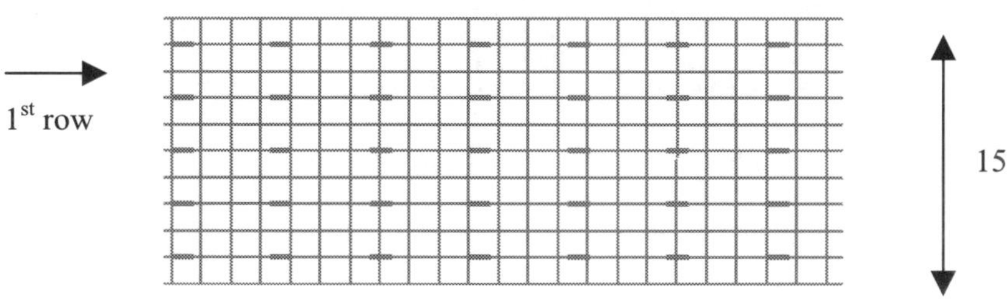

Figure 10: Anodes distribution on the wall of the dock.

The anodes were distributed on the wall of the dock as it is shown in figure 10. When no ship is near the dock the potential distribution is as shown in the Figure 12.

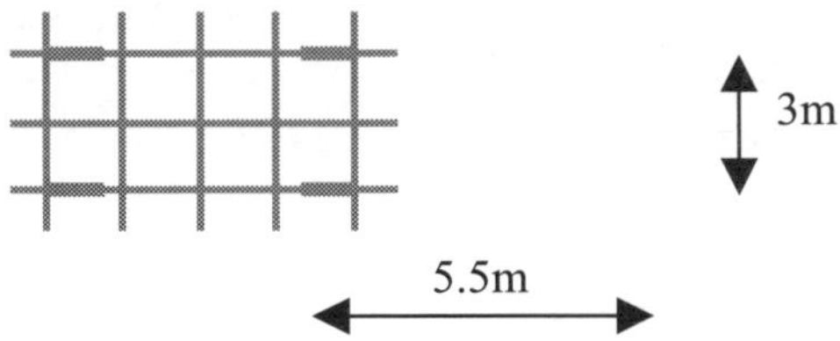

Figure 11: Distance between anodes on the wall.

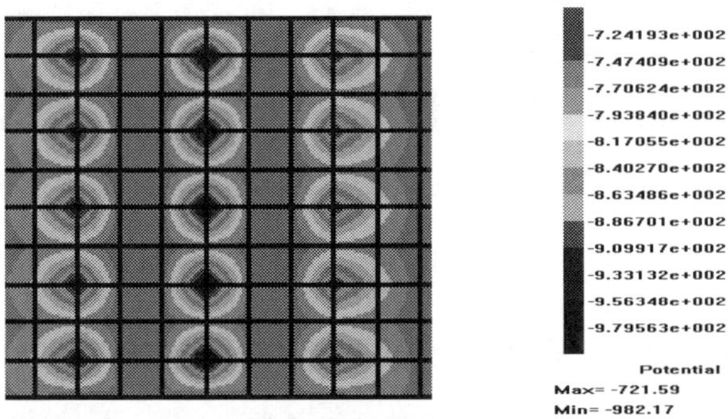

Figure 12: Cathodic protection of the dock when no ships in its surroundings.

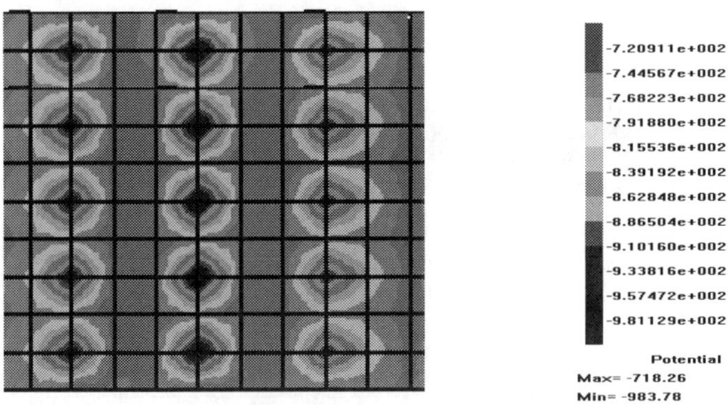

Figure 13: Cathodic protection of the dock when the frigate is berthed.

In addition, the current supplied by anodes on the frigate is very similar to the current supplied when no anodes were placed on the dock. In contrast to the case with no sacrificial anodes, the dock is protected from corrosion and no anodic areas are found.

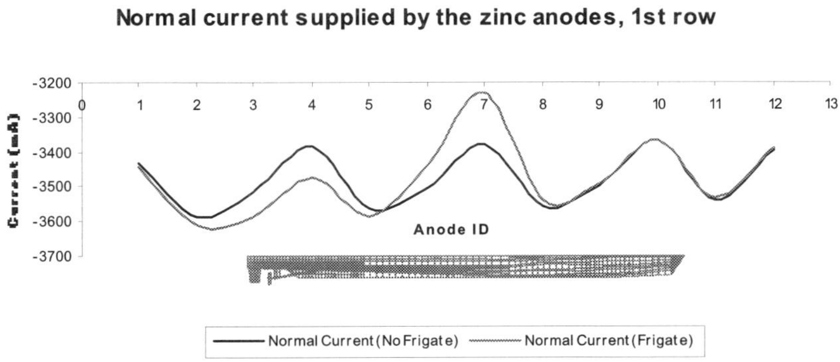

Figure 14: Current supplied by the anodes when there was and there was not a ship berthed (1st row).

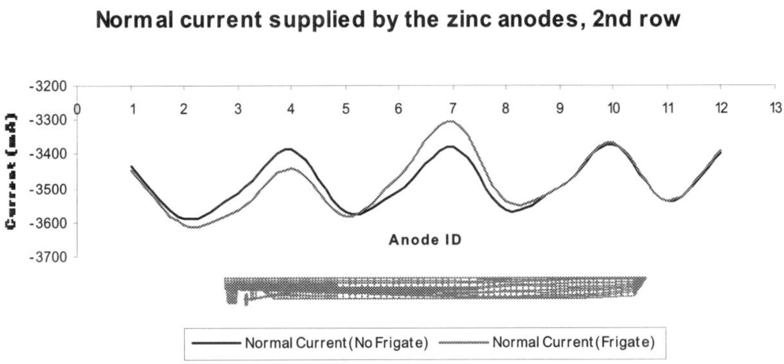

Figure 15: Current supplied by the anodes when there was and there was not ship berthed (2nd row).

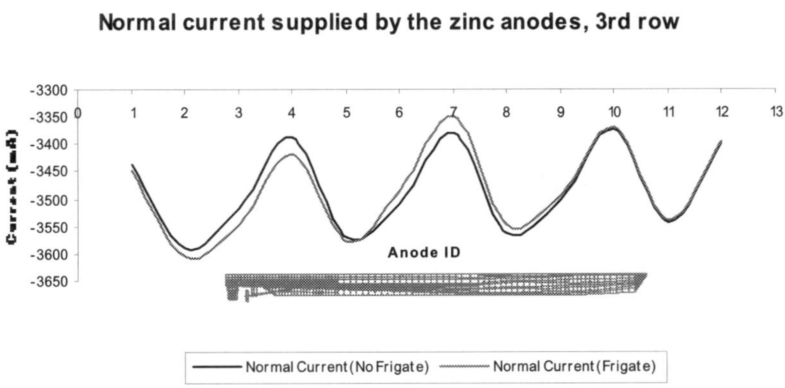

Figure 16: Current supplied by the anodes when there was and there was not ship berthed (3rd row).

The impressed anodes of the frigate still supply current to the dock and the current returns back to the frigate near the propeller position. However, the current returning back from the sacrificial anodes causes a higher consumption of these anodes. Thus, it can be seen in Figure 14–Figure 18 that the current supplied by the sacrificial anodes of the dock near the impressed anodes supply less current to the dock when the frigate is berthed than when there is no ship around. In contrast, the sacrificial anodes near the propeller supply more current to the dock and to the frigate.

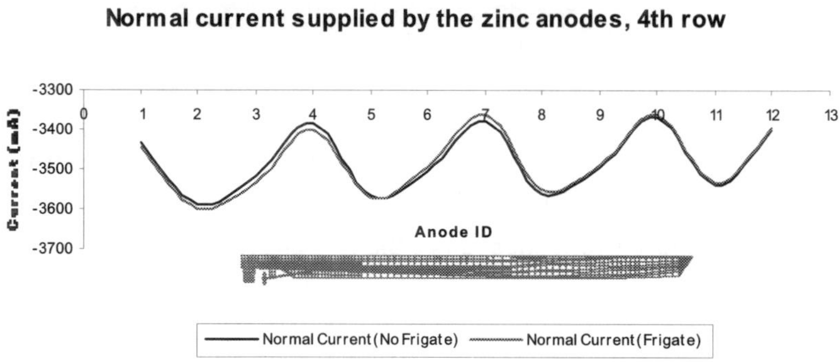

Figure 17: Current supplied by the anodes when there was and there was not ship berthed (4th row).

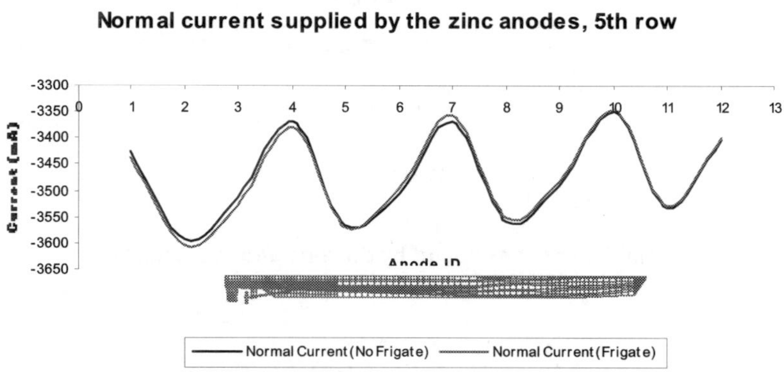

Figure 18: Current supplied by the anodes when there was and there was not a ship berthed (5th row).

8 Conclusions

Quantitative data on the behaviour of the cathodic protection system of a frigate in open sea and its influence/interference on a steel dock has been shown by using computer simulation software. A possible mitigation of the interference by placing a sacrificial cathodic protection system on the walls of the dock was attempted with the results that still there were stray currents coming from the frigate into the dock and from the sacrificial anodes of the dock back to the frigate. Thus, the sacrificial anodes of the frigate placed near the propeller of the frigate will be consumed faster than the rest. In contrast, the sacrificial anodes placed near the impressed anodes of the frigate will last longer since the frigate is providing part of the current necessary to protect the dock. Anode consumption rates were predicted by the computer model under various possible scenarios.

The simulation has been carried out using the BEASY boundary elements package software [1] and the assistance of an optimisation tool was essential to obtain a faster convergence of the problem. It could be observed that 3D model enables the impact of small changes to the geometric design, coating condition, location of anodes etc to be easily assessed.

References

[1] Brebbia C.A, Boundary Element Method for Engineers, Pentech Press. London, 1978.

[2] Danson, D.J. and Warne, M.A. Current Density/Voltage Calculations Using Boundary Element Techniques, Corrosion/83. 1983.

[3] Thomas, E. D., Lucas, K. E., and Parks, A. R., Verification of Physical Scale Modeling with Shipboard Trails, Corrosion 90, Paper 370, National Association of Corrosion Engineers, Houston, TX, 1990.

[4] V.G. DeGiorgi, A. Kee, E.D. Thomas. Characterization accuracy in modelling of corrosion systems. Boundary Elements XV. Vol. 1: Fluid Flow and Computational Aspect. Editors: C.A. Brebbia, J.J. Rencis. CML Publications, Southampton, 1993.

[5] Gartland P.O, Bjoernass F., Osvoll H. Computer modelling of offshore CP systems for 15 years: What have we learnt? Corrosion NACE Expo 99, San Antonio Texas. 1999.

[6] Adey, R.A., Brebbia, C.A., Niku, S.M. Applications of Boundary Elements in Corrosion Engineering, Chapter 3, Topics in Boundary Elements Research ed. Brebbia, C.A. Vol. 7, pp. 34-64, Springer - Verlag, Berlin and New York, 1990.

[7] Roe D. Stommen, Computer modelling of offshore cathodic protection systems utilized in CP monitoring. Offshore Technology Conference, Texas, May 3-6, 1982.

[8] John Morgan. Cathodic Protection. National Association of Corrosion Engineers, NACE. 1987.

[9] J. Trevelyan and H.P. HackThomas, Analysis of stray current corrosion problems using the boundary element method. Boundary Element Technology IX. 1994.

[10] E Santana Diaz, R Adey. A Computational Environment for the Optimisation of CP system Performance and Signatures, Warship CP2001. Shrivingham. UK, 2001.

[11] Adey R A, Hang P Y, Optimum Design of Ship Corrosion Protection Using Computer Simulation, Computers and Ships Conference, The Institute of Marine Engineers, London, UK, May 1999.

[12] Adey R A, Baynham J M W, Design and Optimisation of Cathodic Protection Systems Using Computer Simulation, NACE's Annual Conference "Corrosion 2000", Orlando, Florida, USA, March 2000.

[13] E. Santana Diaz, R Adey, J. Baynham, Optimising the location of anodes in cathodic protection systems, OPTI 2003, May 2003.

[14] K. Kishimoto, K. Amaya, S. Aoki, Optimization of cathodic protection using boundary element method. In Boundary Element Methods. Principles and Applications, ed. M. Tanaka and Q. Du. Pergamon Press, Oxford, pp. 329-338 1990.

Zonal ICCP system control interactions

V. G. DeGiorgi[1], M. Pocock[2], S. A. Wimmer[1] & E. A. Hogan[1]
[1]*Naval Research Laboratory, USA*
[2]*Frazer-Nash Consultancy Ltd., UK*

Abstract

To date much work has been done on modeling shipboard impressed current cathodic protection (ICCP) systems. The emphasis has been on system performance: the systems' ability to maintain the hull at a constant potential value. Reported system design work has dealt mainly with anode number, anode placement, subdividing the overall system into power supply zones (anodes connected to a single power supply with a controlling reference cell) and reference cell placement. As electronic and hardware advances are incorporated into new ship designs the system control algorithms are evolving. In general, system control algorithms can range from simple one point feedback to complex multiple input-multiple-output that subdivide the structure into small regions. Multiple input-multiple-output systems provide the ability to fine tune in order to maintain uniform values throughout the structure. It has long been observed, both experimentally and on shipboard, that even simple ICCP systems with two or three power supply zones show zonal interactions. This work uses a model ship geometry, computational methods and experimental physical scale model results to further the understanding of the interaction between individual power zones on a typical shipboard ICCP system.
Keywords: cathodic protection, ICCP system, zone interaction, control, polarization response.

1 Introduction

Modern ICCP systems are subdivided into zones; each zone contains multiple anodes which are driven by a single power supply. Each zone has its own controller and reference cells that feedback information on surface potential at defined points to the controller. Despite the fact that designers may speak in terms of zones with the implication that there are zonal boundaries the hull is a continuous structure and there are no physical barriers that isolate zones from each other. There is experimental and shipboard experience that tells us that zonal interactions occur. Operators and experimentalists use their developed understanding of ship and system performance to adjust reference cell potentials so that over and under protection does not occur. This is a learned expertise and as such will be difficult to apply to more complex zone systems. Computational modeling focused on zone interactions may provide additional insight to system performance.

Knowledge of ICCP systems, including understanding of the individual components, is necessary in order to understand the issues related to zone interactions in ICCP systems. ICCP systems exploit the electrochemical nature of corrosion and the establishment of cathode-anode regions on the ship. Cathodes (exposed metal regions on the hull, rudder and propeller) are protected regions and anodes are either discrete components of sacrificial metal or electron source points positioned along the underwater hull at specific locations. For a ship at sea, the wetted surface area of the hull and the appendages such as bilge keel, rudder and propeller, are the cathodes, which require CP. When a ship is freshly out of dry-dock, the cathode areas are typically the propellers and small regions on the hull where the coatings may not be adequate (including small breaks in the paint which are referred to as paint holidays). These are the only regions that require protection. The current demand is very low and does not tax the power capability of the system. As the ship spends time in service, the coatings degrade, adsorb moisture and ultimately break down in regions exposing additional metal surfaces. In addition mechanical damage to coatings often occurs with time in service. The result of these increases in exposed metal areas over time is an increase in the power required to maintain the levels of protection required to prevent corrosion. As the coatings age, additional power is required. If the system has been properly designed ample power will be available and power sources are distributed in such a way that a suitable potential is maintained over the entire hull.

A typical ICCP system consists of non-sacrificial noble anodes connected to power supplies, reference cells to monitor hull potential state and a controller to adjust the current output of the anodes. Existing systems depend on simple feedback control algorithms. New systems are not limited to this type of control. The ideal system is designed with anodes located so current is evenly distributed so that a uniform voltage is maintained for all points on the underwater hull. In reality, cathodes created by paint damage, components made of non-similar materials, geometric features and openings in the hull result in a varied profile.

Ship hull condition, and therefore ICCP system performance, is transient in nature. Paint ages with time and environmental exposure. Both damage and protective film formation occur over time. The controller and its control algorithm allow the ICCP system to respond to variations with time and operating conditions.

Anodes, reference cells and power supplies are grouped to form zones. A zone is defined as a single controller that adjusts the output of one or more power supplies through their electrically associated anodes. Reference cells are used to monitor the hull potential and provide information to the control algorithm. Typically only 1 or 2 reference cells are used per zone. Therefore the placement of reference cells so that recorded values are representative of a region of the hull is critical. The size and complexity of modern navy vessels has led to the development of multiple zone systems to adequately protect the diverse geometric areas of the hull. Existing systems predominately use 2 or 3 zones.

Once the concept of zones was incorporated into the ICCP system, the effects of one zone's power input on the reference cell readings of another zone becomes an issue. Reference cells provide information on regions of a hull. However there is no barrier isolating one zone from another. This work is the first in a series that examines zonal interactions in shipboard ICCP systems. The objective is to obtain qualitative information on the degree of interactions that occur. In the present work computational modeling is used to examine the interaction between the zones of a 2-zone system. Boundary element methods have been demonstrated appropriate for shipboard ICCP system evaluation [1, 2] and are used for the computational work. The first step is an examination of the computational solution methodology used. Two different solution approaches, manual iteration and automated computer solution, are used. In addition the solutions are compared with physical scale model experimental results to validate the boundary element model. Once the computational process was validated, a series of zonal influence cases were evaluated. Results are used to develop an understanding of the extent and magnitude of zonal interactions.

2 Ship geometry and boundary element modeling

The geometry used for this study is representative of navy ship hulls. There is one propeller and one rudder located along the port-starboard centerline. Damage is defined so that the pattern is similar to those defined for U S Navy surface ships by the Naval Sea Systems Command.

The ship is outfitted with a 2-zone ICCP system. Anode placement, power supply sizing, and reference cell locations are defined based on established design rules. The ICCP system is not designed using physical scale modeling experimental methods. This system is representative of older systems still in use. The system is not presented as an optimum system. It is similar to existing systems and as such is used as a starting point for qualitatively determining zonal interaction characteristics.

The ship hull boundary element mesh is shown in Figure 1. The mesh was created using the commercial computer code PATRAN [3]. Details of anode, reference cell and power supplies are shown in Table 1. In the current work we focus on one load case; 3% exposed surface area corresponding to newly painted conditions. Exposed metal is limited to docking block areas (steel) and the propeller. Coatings are considered to be perfect dielectric surfaces. Damaged, or cathode, areas are assigned the polarization response of steel. The propeller is assigned the polarization response of Nickel-Aluminum-Bronze (NAB). Material properties are defined as discussed in Ref. [4]. The ship is contained in an infinite volume of seawater with conductivity of 5.69 x 10^{-2} Siemens/m.

Anode input values were defined by the analyst for manual and automated solution methodologies. The set reference cell value for this work is –0.85V verses an Ag/AgCl reference cell, ± 0.05V, the U S Navy standard. The commercial boundary element code FNREMUS [5] was used in this work.

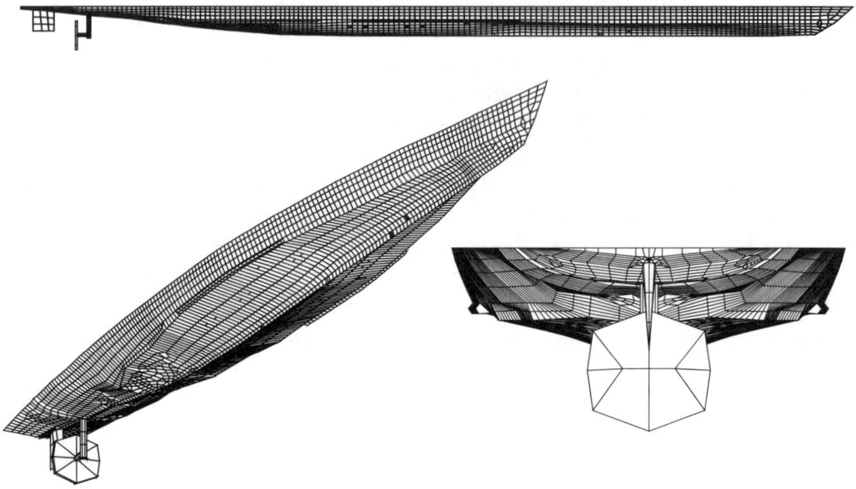

Figure 1: Boundary element mesh.

3 Solution methods

3.1 Manual solution methodology

The manual solution methodology has been used with success in past work at NRL and other institutions. Prior to the development of solution algorithms, such as those documented in [5] and [6], this was the only approach to determine ICCP system parameters. In this method the analyst defines the anode input values, allows the program to numerically solve the system of equations and then compares nodal point values at specific locations with reference cell set point values. In addition to this comparison of nodal values, the total current associated with each power supply is calculated and compared with the maximum allowable power level. In order to be a valid solution, nodal potential values at reference cell locations must be within the defined range for set point values and the total current output for each power supply must be equal to or less than the prescribed maximum for that particular power supply. Input values at anodes can be defined in terms of current, current density or voltage. The input values are adjusted based on calculated results. The number of iterations required to obtain a valid solution is greatly dependent on the skill and expertise of the analyst. A working knowledge of ICCP system performance, either from experimental, ship operations or previous calculations, provides insight in establishing the initial anode values. Once upper and lower bound solutions are obtained, either a simple incremental or more complex mathematical scheme can be used to adjust the anode. This approach is trial and error with incremental corrections to input values to converge on a final valid solution. There is no rigorous mathematical basis for establishment of an obtained solution as an optimum solution.

Table 1: ICCP system details-scale geometry dimensions.

	Frame	Below Waterline (cm)	Off Centerline (cm)	Symmetry
Forward Zone Anodes	49.9	2.3	1.8	Port & Starboard
	189.1	2.8	4.6	Port & Starboard
Forward Zone Reference Cells	103.7	3.3	1.0	Port & Starboard
Aft Zone Anodes	279.4	2.8	4.3	Port & Starboard
	368.8	0.7	3.8	Port
	379.5	0.5	3.8	Starboard
Aft Zone Reference Cell	397.0	0.5	0.2	Starboard

The manual system becomes increasing difficult for the analyst as the number of zones increases. The manual solution method has been successfully applied to 2 and 3 zone systems. These systems used direct feedback between the reference cells and power supplies in which the control algorithm requests an increase or decrease in power directly proportional to the difference in reference cell reading and control set point.

As ICCP systems evolve and make use of more complex control algorithms the use of the manual iteration techniques is limited. While this method can still be used to obtain a snapshot in time of ICCP system performance, it is hard to imagine applying this approach to complex control systems that change over time and conditions.

3.2 Automated solution control algorithm

In essence an automated solution algorithm mimics the action of an ICCP controller, and automates the manual iterative process described above. It is also an iterative process. The anode current for each iteration is determined from the difference between the reference cell's set potential and achieved potential at the previous iteration. The relationship between potential and current can be a simple linear gain, a Proportional-Integral-Derivative (PID) type control or a more complex algorithm, and can take account of maximum supply currents or other physical constraints of the ICCP system. At each iteration the currents at all anodes are adjusted to minimize the difference between set and achieved potential at each reference cell.

The system may comprise several ICCP zones, with each zone comprising one or more anodes and one or more reference cells. A separate algorithm is used for each zone, and the control of each acts independently of the others to determine the current of its anode(s) from the potential at its reference cell(s). As in the real world, the zones interact with each other since the anode current from one zone will influence the potential at the reference cell of other zones. The automated solution algorithm will adjust each zone's currents taking account of these interactions.

The main advantage of the automated system is that human intervention is not required after each iteration to determine the next iteration's anode currents. The solution process becomes a simple matter of setting up the ICCP parameters and letting the process run to convergence. When complex multi-zone systems are being simulated the benefits of an automated system are greatly increased.

An automated solution control algorithm has been implemented in the FNREMUS code [5].

4 Results

Results are presented for the ship minimum damage station condition for four zonal operation combinations. These four combinations are (1) normal operations conditions with both zones active, (2) forward zone off without additional power, (3) forward zone off with additional power so that the aft zone reference cell is at the set potential and (4) forward zone off with additional power so that both reference cells are at the set potential. Each of these combinations relates to a real world operating scenario.

In normal operating conditions both forward and aft zones are fully functional. For this case computational results are compared with experimental physical scale model generated results. Physical scale modeling (PSM) is physics based experimental process used to design and evaluate ICCP systems. Details of this approach can be found in Ref. [7]. Key system parameters are shown in Table 2. Even though current totals shown good agreement there is significant difference in zonal inputs between computational and experimental results. This is an indication that there are multiple combinations of zonal input values that result in similar overall system performance parameters. There have been indications that this may occur in past work and is an area for future research.

Two definitions of the material polarization curves have been used. Both curves are based on the same experimental data. For the manual method, the electrochemical behavior of the material is represented by a piece-wise linear curve. For the automatic algorithm, a smooth curve based on Tafel and Butler-Volmer electrochemical equations is used. The smooth curve is preferred for the automated algorithm as the discontinuities in the piece-wise curve may lead to instabilities; however both representations produce very similar results as shown in Table 2.

Table 2: System parameters for normal operations with both zones active.

	PSM	Manual Iterations	Auto Piece-wise	Auto Smooth Curve
Fore Anodes (mA)	1.21	0.736	0.154	0.161
Aft Anodes (mA)	0.76	1.220	1.772	1.812
Total	1.97	1.956	1.926	1.973
Docking Blocks (mA)	0.89	0.902	0.875	0.889
Propeller (mA)	1.15	1.050	1.051	1.084
Total	2.04	1.952	1.926	1.973
Forward Reference Electrode (mV)	850	864	851	851
Aft Reference Electrode (mV)	850	856	850	850

Turning off the forward zone without additional power input gives the analysts a snapshot of what occurs when one zone fails. The condition as modeled is prior to the remaining system being adjusted to compensate for the loss. If there is no zonal interaction, the resulting potential map would show under protection only for that region associated with the forward zone. As can be seen in Figure 2 and Table 3, under protection occurs for a significant about of the hull demonstrating that there are interactions between zones. As seen in Table 3, both computational solution methods result in almost identical results.

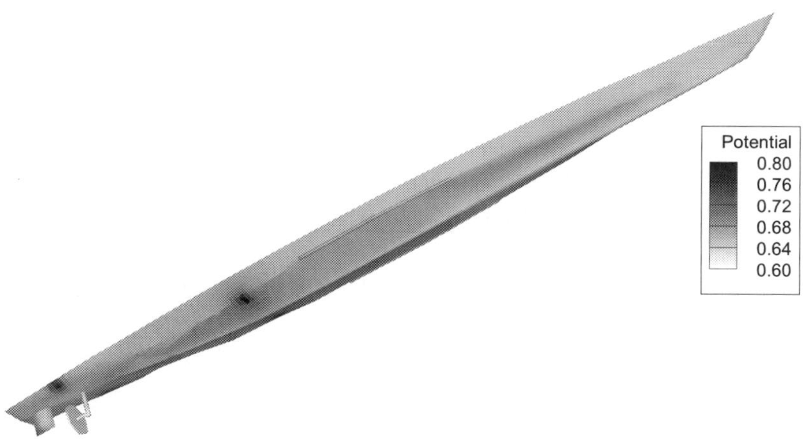

Figure 2: Contour plot of potential (V) of the solution via manual iterations for forward zone off without additional power.

WIT Transactions on Engineering Sciences, Vol 48, © 2005 WIT Press
www.witpress.com, ISSN 1743-3533 (on-line)

Table 3: System parameters for forward zone off without additional power.

	Manual Iterations	Auto Smooth Curve
Fore Anodes (mA)	OFF	OFF
Aft Anodes (mA)	1.22	1.812
Total	1.22	1.812
Docking Blocks (mA)	0.265	0.749
Propeller (mA)	0.952	1.063
Total	1.217	1.812
Forward Reference Electrode (mV)	649	799
Aft Reference Electrode (mV)	636	796

Table 4: System parameters for forward zone off with additional power so that the aft zone reference cell is at the set potential.

	Manual Iterations	Auto Tafel
Fore Anodes (mA)	OFF	OFF
Aft Anodes (mA)	1.92	1.969
Total	1.92	1.969
Docking Blocks (mA)	0.868	0.884
Propeller (mA)	1.052	1.085
Total	1.920	1.969
Forward Reference Electrode (mV)	846	847
Aft Reference Electrode (mV)	850	849

Additional power to bring the working zone's reference cell into the set potential range is another method to quantify the extent of zonal interactions. If the degree of interaction is low then there would be less response in the non-working reference cell than there is the working zone's reference cell. As can be seen in Table 4, both the aft and forward zone reference cells show significant changes in values as additional power is added to the system from the aft zone. This shows a definite connection between the responses of the regions covered by the two zones.

Table 5: System parameters for forward zone off with additional power so that both reference cells are at the set potential.

	Manual Iterations	Auto Smooth Curve
Fore Anodes (mA)	OFF	OFF
Aft Anodes (mA)	1.936	1.979
Total	1.936	1.979
Docking Blocks (mA)	0.882	0.893
Propeller (mA)	1.054	1.086
Total	1.936	1.979
Forward Reference Electrode (mV)	850	851
Aft Reference Electrode (mV)	855	854

Figure 3: Contour plot of potential (V) of the solution via manual iterations for forward zone off with additional power so that both reference cells are at the set potential.

Finally, the ability of a single working zone to provide sufficient impressed current so that both reference cells are within the set potential range is demonstrated. As can be seen in Table 5, there is only a small increase in power required to achieve this condition over the last condition. This is another indication of the high level of interactions between zones for this ship hull and ICCP system design. The model ship geometry and ICCP system provide coverage even with the loss of power to one zone as can be seen in Figure 3.

The ability for sufficient coverage when all zones are not operational is conditional on power ratings for each zone and hull geometry. The ability to maintain protection without significant over-protection in some areas is highly dependent on geometric considerations (hull geometry, anode placement, reference cell placement) and should be seen as a fortuitous circumstance when it occurs. It is far too costly to over design systems with the redundancy required to insure this capability. The small variation between the last two zone combinations in this work should not be seen as typical.

5 Summary

In this work we have shown computational results that clearly indicate zonal interactions, as described by ship operations, do exist and that these interactions have a very real impact on system performance. The work presented is preliminary in nature and is the first step in a detailed analysis planned to quantify zonal interactions. The objective of the presented and planned work is twofold; to better understand the physical phenomenon associated with the operation of ICCP systems and to develop new computational and analytical tools that take into account real world operations for the improved design of ICCP systems.

Acknowledgements

The support of Dr. A. I. Kaznoff and Mr. E. D. Thomas, Naval Sea Systems Command, is gratefully acknowledged.

References

[1] Adey, R.A. & Niku, S.M., "Computer modeling of corrosion using boundary element method," ASTM STP 1154, American Society Testing and Materials, 248-264, 1992.

[2] DeGiorgi, V.G., Hogan, E., Lucas, K.E. & Wimmer, S. A., "Computational Modeling of Shipboard ICCP Systems," *J. Corrosion Science and Engineering*, Vol. 4, Paper 3, 2003.

[3] MSC Software Corp., MSC Patran 2001 User's Manual, 2001.

[4] DeGiorgi, V.G., "Evaluation of Perfect Paint Assumptions in Modeling of Cathodic Protection Systems," *Engineering Analysis with Boundary Elements*, 26/5, 435-445, 2002.

[5] Frazer-Nash Consultancy, FNREMUS Detailed Modeller User Guide, FNC 5421/21133R, Issue 1, 2000.

[6] Diaz, E.S. & Adey, R. "Predicting the hull state from information of potential measurements of the hull," Proceedings 11th International Warship Cathodic Protection Symposium and Equipment Exhibition, Defence Academy of the UK, 2003.

[7] DeGiorgi, V.G., Hogan, E., Lucas, K.E. & Wimmer, S.A., "Shipboard Impressed Current Cathodic Protection System," chapter in BEM and Corrosion, WIT Press, In Press.

Optimum design of cathodic protection using the Boundary Element Method

K. Amaya[1] & S. Aoki[2]
[1] *Tokyo Institute of Technology, Ookayama, Meguro, Tokyo, Japan*
[2] *Toyo University, Kujirai, Kawagoe, Saitama, Japan*

Abstract

Optimization of cathodic protection systems for a pipeline and a ship was performed by using the boundary element method (BEM). For a pipeline, the location and impressed current of electrodes in a cathodic protection system were optimized by minimizing the electric power necessary to keep the potential on the metal surface below a critical value. The non-uniformity of soil conductivity and the resistance of a long pipeline were taken into account. For a ship, an inverse problem to estimate the potential distribution on the whole surface of the hull from the electric potential data measured with several sensors was solved at first. Then, the optimum current to be impressed to each electrode located on the hull was determined by using the results of the inverse analysis.

1 Introduction

Protecting structures, such as a pipeline and a ship, from corrosion is one of the most important problems in engineering [1]. To prevent corrosion of a pipeline, its outer surface is painted and also cathodic protection is performed by impressing current into surrounding soil from a finite number of electrodes. It is necessary to optimize the location and current to be impressed to each electrode.

Application of the boundary element method to the optimization has been studied intensively [2]-[6]. Special consideration is necessary for a pipeline. because it is so long that the non-uniformity of the electric conductivity of soil must be taken into account, and also the electric resistance of the pipeline can not be neglected [7, 8].

The cathodic protection is performed also for a ship by impressing current into surrounding sea from several electrodes on its hull. Some sensors for monitoring

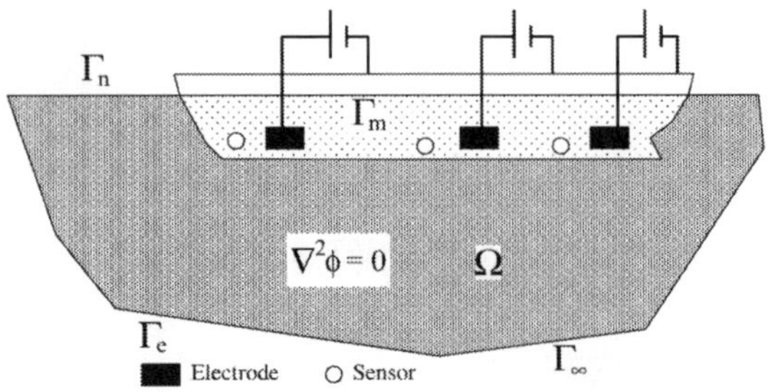

Figure 1: Impressed current CP for ship.

potentials are located at several points on the hull. During navigation, it is necessary to estimate the potential on the whole surface of hull from the potential data measured by a small number of sensors and determine the optimal current to be impressed to each electrode [8, 9].

In this paper, after reviewing the electrochemical aspects of corrosion, a mathematical model, boundary element formulation, solution methods are presented with a few example problems [7]-[9].

2 Electrochemical background

We consider a typical metal M in an electrolyte. Both anodic and cathodic reactions occur simultaneously on the metal surface according to:

Anodic reaction :

$$\mathrm{M} \rightarrow \mathrm{M}^{n+} + \mathrm{ne}^- \tag{1}$$

Cathodic reaction:

$$\frac{1}{2}\mathrm{O_2} + \mathrm{H_2O} + 2\mathrm{e}^- \rightarrow 2\mathrm{OH}^- \tag{2}$$

Due to the reactions an electric current flow occurs. The electric current density i versus electrical potential to a reference electrode (e.g., saturated calomel electrode) E curve is called a polarization curve. The relationships between the current density i and the potential E for anodic and cathodic reactions are not obtained individually but nominal relationship $E = f(i)$,i.e., the algebraic sum of the currents for the two reactions versus E curve is measured.

In the natural state, the reactions become in equilibrium, and anodic current flows from the anode to the cathode. Corrosion rate is proportional to this anodic current density. It is possible to supress the anodic current density by impressing current from external power supply and reducing the potential of the metal to the critical value E_p. This method is called cathodic protection.

3 Ship

It is necessary to estimate the potential on the whole surface of hull from the potential data measured by some sensors on the hull, and determine the optimal currents to be impressed to each electrode.

3.1 Basic equations

Let us assume that the region occupied with sea water, Ω, is surrounded by the surface of the hull, Γ_m, the surfaces of electrodes, Γ_e ($e = 1, 2, \ldots, N$; $N =$ total number of electrodes), the surface of sea water, Γ_n and infinite boundary, Γ_∞ as shown in figure 1. The potential, ϕ, satisfies the following Laplace equation in Ω;

$$\nabla^2 \phi = 0 \quad \text{in } \Omega \tag{3}$$

and is given at the location of sensors, $\boldsymbol{x}_s$ ($s = 1, 2, \ldots, M$; $M =$ total number of sensors);

$$\phi = \phi_s \quad \text{at } \boldsymbol{x}_s \tag{4}$$

where $\boldsymbol{x}_s$ is the location of a sensor and ϕ_s is the (known) value of potential measured by the sensor s. The ϕ is defined as $\phi = -E$, where E is the potential with respect to some reference electrode, e.g., SCE.

The current density, i ($\equiv \kappa \partial\phi/\partial n$; $\kappa =$ conductivity of sea water, $\partial/\partial n =$ outward normal derivative) satisfies the following equations;

$$i = 0 \quad \text{on } \Gamma_n \tag{5}$$

$$i = 0 \quad \text{on } \Gamma_\infty \tag{6}$$

$$i = i_e \quad \text{on } \Gamma_e \tag{7}$$

where i_e ($e = 1, 2, \ldots, N$; $N =$ total number of electrodes) is the current density on the electrode e. Let $\boldsymbol{Q}$ represent a vector, the components of which are given with i_e.

3.2 Corrosion rate identification

Let us consider an inverse problem where the potential distribution on the surface of hull is estimated from the potential data measured by several sensors. We will use the boundary element method. If we chose the nodal values of potential of each element as the unknown parameters, the number of potential data would be too small compared with that of unknown parameters. Even if the number of data could be increased enough to solve the inverse problem, the solution would be abnormally oscillated due to the ill-posedness [11].

Hence, let us make use of the a-priori information that ϕ and i on Γ_m must satisfy the polarization characteristics. At first, we will approximate the polarization curve by a function, e.g., Tafel's expression [1];

$$-\phi = f_m(i; \alpha_j) \quad \text{on } \Gamma_m \tag{8}$$

where α_j ($j = 1, 2, \ldots, L$; $L =$ total number of parameters) is a parameter in the function. Then, the several parameters in the function are estimated by an inverse analysis. After obtaining the polarization characteristics, the potential distribution is easily calculated by a direct analysis.

Following the usual boundary element formulation, we obtain the following equations from (3) [3];

$$\kappa[H]\begin{Bmatrix} \phi_\infty \\ \phi_e \\ -f_m(i_m; \alpha_j) \end{Bmatrix} - [G]\begin{Bmatrix} 0 \\ i_e \\ i_m \end{Bmatrix} = 0 \tag{9}$$

By assuming a set of initial values of the unknown parameters, α_j, we solve a direct problem, i.e., solve (9), at first and then minimize the following cost function;

$$cost(\alpha_j) = \sum_{s=1}^{M} (\phi_s(est) - \phi_s)^2 \tag{10}$$

where $\phi_s(est)$ is estimated value of potential at the location of sensor s. Calculations are repeated by modifying the parameters employing a minimizing technique, e.g., the conjugate gradient method, until the value of cost function, $cost(\alpha_j)$, becomes less than the measuring error, ε.

Even if the above procedure is followed, the solution for α_j sometimes depends on the assumed initial values. By varying initial values widely, we can obtain the possible solution area, Π. In order to reduce the area, Π, we use the potential data obtained under different sets of i_e (impressed current density on each electrode) as well as the original data. Furthermore, we utilize fuzzy a-priori information expressed as membership functions [1]. For example, a-priori information are such that if the hull is made of a low alloy steel, then the parameters of polarization characteristics curve must be within some range. Let $F_i (i = 1, \cdots n;\ n =$ the number of a-priori information) denote the fuzzy set expressed by the membership function. The solution can be obtained as the center of gravity of a set, $\Pi \cap F_1 \cap F_2 \cdots \cap F_n$.

3.3 Optimization of impressed currents

The power necessary for cathodic protection is give by the following equation;

$$P(\boldsymbol{Q}) = \sum_{e=1}^{N} \int_{\Gamma_e} \{\phi_e(\boldsymbol{Q}) + f_e(i_e(\boldsymbol{Q}))\} i_e(\boldsymbol{Q}) d\Gamma \tag{11}$$

where the polarization characteristics of the electrode e is given by a known function, $-\phi = f_e(i)$, and the components of a vector $\boldsymbol{Q}$ are i_e $(e = 1, 2, \cdots, N)$. It is necessary to satisfy the following protection condition on every part of Γ_m;

$$\phi \geq \phi_p \quad (\text{i.e., } E \leq E_p) \quad \text{on } \Gamma_m \tag{12}$$

where $E_p(= -\phi_p)$ is the critical value of potential for protection.

The problem is to find the vector $\boldsymbol{Q}$ which minimizes P in (11) under the condition (12). It is, however, necessary for ϕ and i in (11) and (12) to satisfy (9), where the α_j has already been obtained in the last section. Since α_j changes during navigation, it is necessary to determine effectively the optimal current impressed to each electrode, corresponding to the change of α_j.

To do this, we combine (11) and (12) by defining a new cost function, $P^*(\boldsymbol{Q})$ [8];

$$P^*(\boldsymbol{Q}) = P(\boldsymbol{Q}) + \nu \sum_{k=1}^{K} h_k^2(\boldsymbol{Q}) \tag{13}$$

Here, ν is a penalty number (large positive number) and the function h_k is defined on Γ_m as

$$h_k = (-\phi_k + \phi_p)\, u\,(-\phi_k + \phi_p) \quad (k = 1, 2, \ldots, K) \tag{14}$$

where $u(\cdot)$ is the unit step function, ϕ_k is the potential at the nodal point k and K is the total number of nodes on Γ_m. If the protection condition (12) is not satisfied, $P^*(\boldsymbol{Q})$ takes a large value.

Thus, it becomes possible to determine effectively the optimal currents impressed to each electrode by minimize $P^*(\boldsymbol{Q})$. The procedure is as follows;

1. Assume $\boldsymbol{Q}$.
2. Solve (9) for ϕ and i on Γ_e by using the polarization characteristics obtained in the last section.
3. Evaluate (13) and modify $\boldsymbol{Q}$ using a minimizing technique, e.g., the conjugate gradient method.
4. Return to 2. and repeat the procedure until convergence is reached.

3.4 Numerical example

We consider a ship which has six electrodes and six sensors as shown in figure 3. At first, a direct analysis was performed for the case where the current densities impressed to the electrodes on the left side hull were 0.04, 0.04 and 0.4 [A/m^2] (from bow to stern), and those on the right side hull were 0.01, 0.01 and 0.01 [A/m^2] (from bow to stern). It was assumed that the hull was made of painted low alloy steel plates and their polarization curves are given by

$$\phi = -0.600 \sinh^{-1}(200i) + 0.650 \tag{15}$$

where the units of ϕ and i are [V] and [A/m^2], respectively.

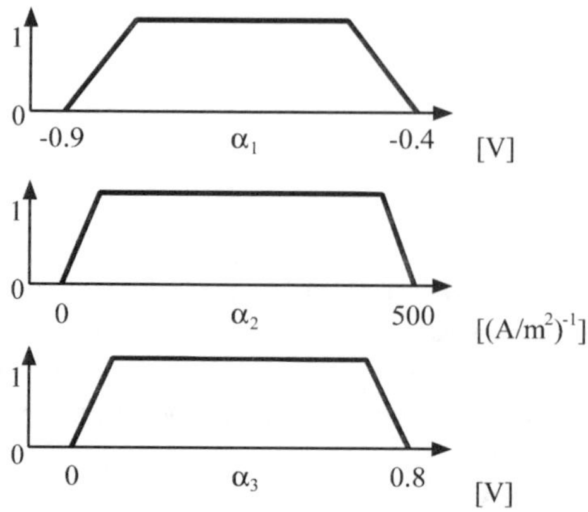

Figure 2: Fuzzy a-priori information about polarization curve.

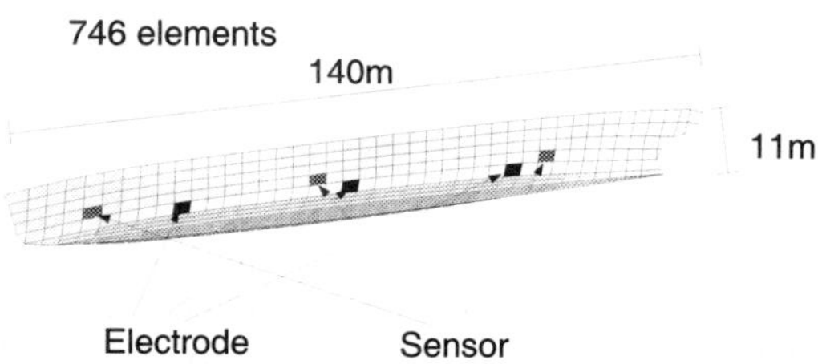

Figure 3: Locations of electrodes and sensors.

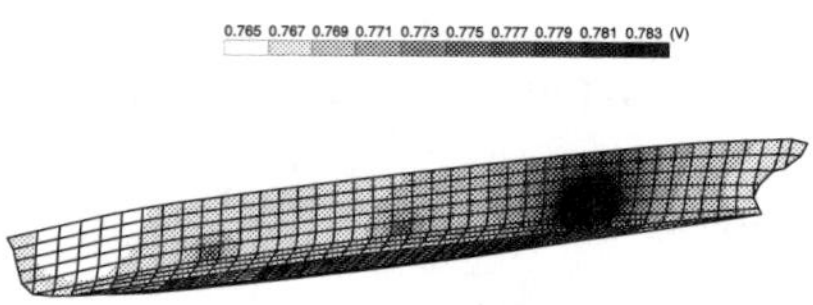

Figure 4: Calculated potential distribution on ship hull.

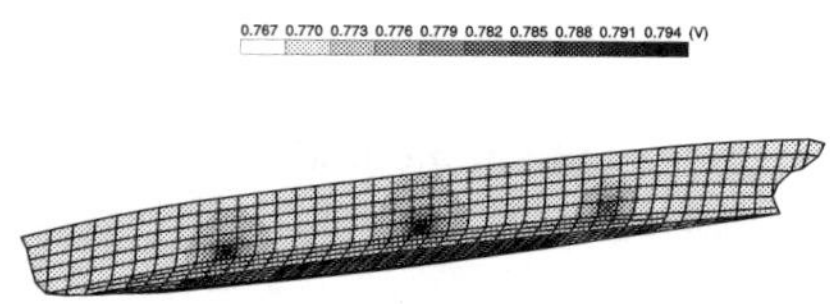

Figure 5: Potential distribution after optimization.

Direct analysis was performed by employing 764 constant elements. The result on potential distribution is shown in Figure 4. The potential values at the location of the sensors were rounded off to three significant figures to take account of the measurement accuracy, and were used as the input data in the following inverse analysis.

In the inverse analysis, the polarization curves were estimated at first. It wa s assumed that the polarization curve of the painted hull is represented in the following form;

$$\phi = -\alpha_1 \sinh^{-1}(\alpha_2 i) + \alpha_3 \tag{16}$$

where $\alpha_j (j = 1, 2, 3)$ are the parameter to be estimated. We used a-priori information about polarization characteristics, the membership functions of which were shown in figure 2.

After estimating these parameters, the potential distribution was calculated by a direct analysis using the estimated polarization characteristics. The estimated potential distribution agreed well with the exact solusion which was shown in Figure 4. It is observed in Figure 4 that the protection condition ($\phi \geq \phi_p = 0.77V$) is not satisfied on the white part of the hull. This is attributable to the improper values of impressed currents to the electrodes.

Optimization of the impressed currents was carried out by using the method described in the preceding chapter. The potential distribution due to the optimized impressed current is shown in Figure 5. It is found that the protection condition is satisfied everywhere on the hull.

4 Underground pipeline

4.1 Basic equations

The pipeline is long that the electrode conductivity of soil can not be assumed to be uniform, but varies gradually as shown in Figure 6 The current density vector $\boldsymbol{i}(\boldsymbol{x})$ [A/m^2] in soil is defined as:

$$\boldsymbol{i}(\boldsymbol{x}) = -\kappa(\boldsymbol{x})\text{grad}\phi(\boldsymbol{x}) \tag{17}$$

where $\kappa(\boldsymbol{x})[\Omega^{-1}m^{-1}]$ is electrical conductivity at a point $\boldsymbol{x}$ in the domain Ω and $\phi(\boldsymbol{x})$ [V] is an electrical potential at $\boldsymbol{x}$.

The electrical potential within the soil, $\phi(\boldsymbol{x})$, obeys the following governing equation:

$$-\text{div}\left\{\kappa(\boldsymbol{x})\text{grad}\phi(\boldsymbol{x})\right\} = \sum_{i=1}^{m_e} I_{e_i}\delta(\boldsymbol{x} - \boldsymbol{x}_{\boldsymbol{e_i}}) \tag{18}$$

where $\delta()$ the delta function, m_e the total number of electrodes, I_{e_i} the impressed currents and $\boldsymbol{x}_{\boldsymbol{e_i}}$ the location of the ith electrodes. It is assumed that the electrodes are so small compared with the size of Ω to be regarded as points.

The pipeline is so long that its internal resistance can not be neglected, and hence the potential in pipeline $\phi = m0$ depends on location $\boldsymbol{x}$. The polarization curve $E = f(i)$ also depends on location $\boldsymbol{x}$. Therefore the boundary condition is described as follows:

$$-\{\phi(\boldsymbol{x}) - \phi_{m0}(\boldsymbol{x})\} = f(i, \boldsymbol{x}) \quad \text{on } \Gamma_m \tag{19}$$

The protective condition is written as:

$$-\{\phi(\boldsymbol{x}) - \phi_{m0}(\boldsymbol{x})\} \leq E_p(\boldsymbol{x}) \tag{20}$$

The optimal locations and impressed currents are to be determined in such a way that the sum of the necessary power to achieve the complete protection and the cost to lay electrodes underground is minimized. We use the following objective function:

$$P(I_e, \boldsymbol{x}_e) = \sum_{i=1}^{m_e} \left(\int_{\Gamma_{e_i}} \{\phi_{e_i} + f_{e_i}(i_{e_i})\} i_{e_i} d\Gamma + k_{z_e} {z_{e_i}}^2 \right) \tag{21}$$

where $\boldsymbol{I}_e, \boldsymbol{x}_e$ are the vectors, the components of which are I_{ei} and $\boldsymbol{x}_{ei}$, respectively. The Γ_{e_i} is the boundary which is radius of ε distant from electrode e_i (ε is very small compared with the size of Γ), ϕ_{e_i}, i_{e_i} and $f_{e_i}(i_{e_i})$ are the potential, the current density and the polarization curve of electrode e_i respectively, k_{z_e} the coefficient of cost to lay the electrode underground, and z_e is the depth of the electrode.

The problem considered here is to find the locations and the impressed currents of electrode that minimize (21) under the conditions (19) and (20).

4.2 Resistance of pipeline

We discretize the pipeline to n nodes. Because of the preservation of the electrical charge, in a jth element of pipeline, we assume the following equation :

$$i_{m0j} S_p - i_{m0j+1} S_p = S_j \frac{i_j + i_{j+1}}{2} + c_j(I_{e_k}) \tag{22}$$

where S_p is the cross section area of pipeline, S_j is the longitudinal area of the jth element, i_j and i_{m0j} are the radial and the axial current density of the jth node respectively. The function $c_j(I_{e_k})(k = 1 \sim m_e)$ is defined as the following:

$$c_j(I_{e_k}) = \begin{cases} I_{e_k}, & \text{if the electrode } e_k \text{ connects} \\ & \text{the } j\text{th element} \\ 0, & \text{if the electrode } e_k \text{ does not connect} \\ & \text{the } j\text{th element} \end{cases} \tag{23}$$

The relations between the radial current density and the axial current density of whole elements are expressed as the following combined form:

$$[A]\{i_{m0}\} = [B]\{i\} + \{I_e\} \tag{24}$$

where $\{i\}$ and $\{i_{m0}\}$ are the vector of the radial and the axial current density respectively, and $\{I_e\}$ is the vector which consists of $c_j(I_{e_k})$.

Following equation can be assumed by the Ohm's law.

$$i_{m0} = -\frac{1}{\rho} \frac{d\phi_{m0}}{dx} \tag{25}$$

where ρ $[\Omega \cdot \mathrm{m}]$ is the resistivity of the internal pipeline.

In each element of the pipeline, assuming the distribution of the axial current density $\{i_{m0}\}$ is linear, the axial potential of the surface pipeline is expressed as:

$$\phi_{m0j} = -\rho \int_{x_{j-1}}^{x_j} \left(\frac{x - x_{j-1}}{x_j - x_{j-1}} i_{m0j} + \frac{x_j - x}{x_j - x_{j-1}} i_{m0j-1} \right) dx + \phi_{m0j-1} \quad (26)$$

where x_j is the location of jth node. Accordingly,

$$\phi_{m0j} - \phi_{m0j-1} = -\frac{\rho}{2}(x_j - x_{j-1})(i_{m0j} + i_{m0j-1}) \quad (27)$$

The base potential is the potential on the nth node.

$$\phi_{m0n} = 0.0 \quad (28)$$

The relations between the axial current density and the internal potential of pipeline are expressed as:

$$[C]\{\phi_{m0}\} = [D]\{i_{m0}\} \quad (29)$$

where $\{\phi_{m0}\}$ the vector of the internal potential of pipeline.

We obtain the relation between the radial current density and the internal potential of pipeline by (24) and (29), as follows:

$$\{\phi_{m0}\} = [E]\{i\} + [F]\{I_e\} \quad (30)$$

4.3 Boundary element formulation

Let us assume the conductivity for the corrosion domain is $\kappa(\boldsymbol{x}) = Kx + K_0$ where K and K_0 are constants and x is the x-component of $\boldsymbol{x}$. The boundary integral equation is expressed as:

$$c(\boldsymbol{y})\phi(\boldsymbol{y}) = \int_\Gamma \{\phi^*(\boldsymbol{x}, \boldsymbol{y})i(\boldsymbol{x}) - i^*(\boldsymbol{x}, \boldsymbol{y})\phi(\boldsymbol{y})\}d\Gamma + \sum_{i=1}^{m_e} I_{e_i}\phi^*(\boldsymbol{x}_{\boldsymbol{e_i}}, \boldsymbol{y}) \quad (31)$$

where Γ is the boundary of the domain under consideration, $\boldsymbol{x} = (x, y, z)^T$ is the position vector of observation point, and $\boldsymbol{y} = (x_s, y_s, z_s)^T$ is the position vector of source point.

In this case, the fundamental solution is given by the following equation [10]:

$$\phi^*(\boldsymbol{x}, \boldsymbol{y}) = \frac{1}{2\pi K r\sqrt{r^2 + 4(x + K_0/K)(x_s + K_0/K)}} \quad (32)$$

$i^*(\boldsymbol{x}, \boldsymbol{y})$ is expressed as:

$$i^*(\boldsymbol{x}, \boldsymbol{y}) = \kappa(\boldsymbol{x})\frac{\partial \phi^*(\boldsymbol{x}, \boldsymbol{y})}{\partial n} \quad (33)$$

where n is the component of the unit outward normal vector at $\boldsymbol{y} \in \Gamma$ and $\frac{\partial}{\partial n}$ denotes the partial derivative along the same normal vector.

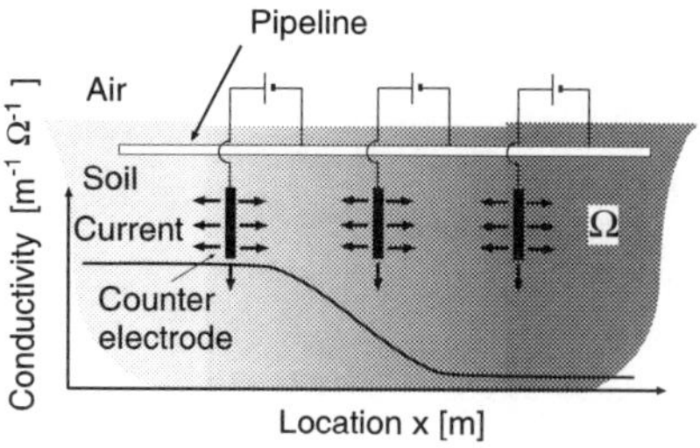

Figure 6: Cathodic protection system for pipeline.

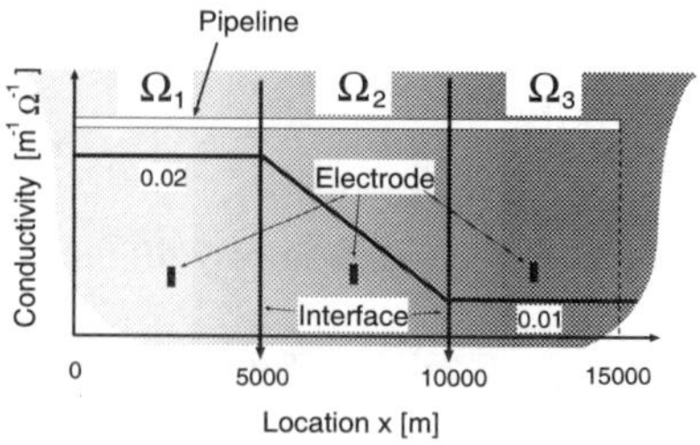

Figure 7: Example problem impressed current CP for pipeline.

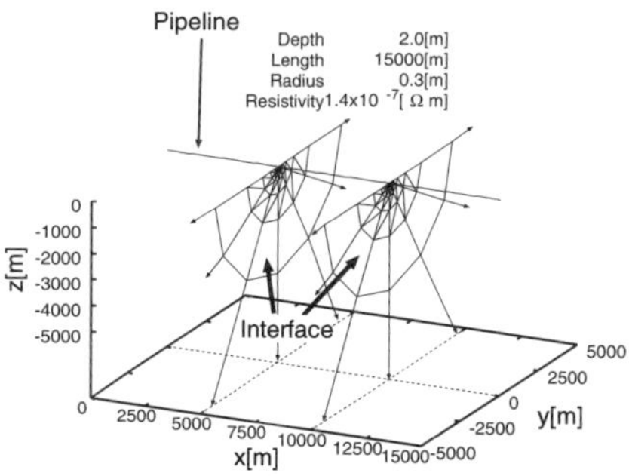

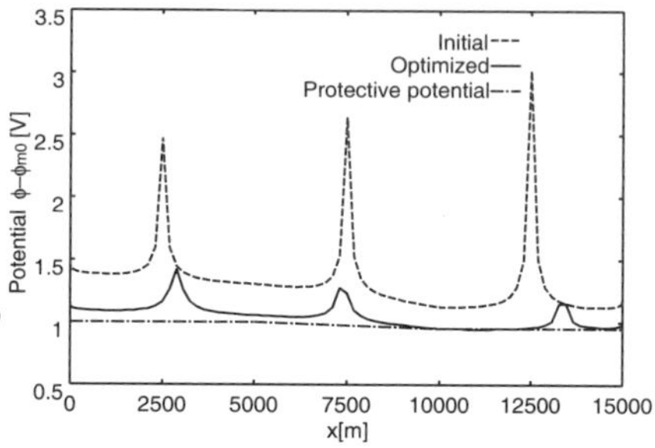

Figure 8: Boundary element mesh for underground pipeline.

Figure 9: Potential distribution on pipeline.

By applying the proposed method in the above section, we can optimize the location of electrodes and impressed current to each electrode.

4.4 Numerical example

An example is shown in Figure 7 in which a pipeline is laid in soil with non-uniform electric conductivity. Let $\boldsymbol{x} = (x, y, z)^T$. The design variables are the locations x_{ei}, x_{ei} and the currentsI_{ei} of electrode ($i = 1, 2, 3$. i.e., $N = 3$), assuming the electrode locations $y_{ei} = 0.0$.

The whole domain is divided into three subdomains $\Omega_1, \Omega_2, \Omega_3$. The electric conductivities in these sub-domains are shown in Figure 7, and the boundary element mesh is shown in Figure 8.

The data in Table 1 was used for the initial design. The electric power was 265.0[W]. Other input data should be referred to [7, 8].

Figure 9 shows the distribution of potential before and after optimization. It is found that the pipeline was over-protected before optimization, while after the optimization the whole pipeline is economically protected. Table 2 shows the optimization results. The electric power was reduced from 265.0 [W] to 83.1 [W].

Table 1: Initial design.

	$x_e[m]$	$z_e[m]$	$I_e[A]$
1	2.5×10^3	-55.0	10.0
2	7.5×10^3	-55.0	10.0
3	12.5×10^3	-55.0	10.0

Table 2: Optimal design.

	$x_e[m]$	$z_e[m]$	$I_e[A]$
1	2.88×10^3	-141.4	8.74
2	7.39×10^3	-107.5	5.00
3	13.36×10^3	-73.7	2.56

5 Conclusions

Optimization of the locations and impressed currents of electrodes in a cathodic protection system for a long pipeline buried in soil with non-uniform electric conductivity was performed by minimizing the cost to lay the electrode underground plus the electric power necessary to keep the potential on the metal surface below a critical value. In the optimization, the multi-region boundary element method (BEM) was applied, and the fundamental solution suitable to the linear change of the electric conductivity was used. The resistance of a long pipeline was also taken into account.

An inverse problem accompanied with performing optimal cathodic protection for a ship was solved, i.e., the potential distribution on the hull was estimated from the potential data at several fixed sensors. To remove the ill-posedness of this inverse problem, the polarization characteristics of the painted hull were estimated at first and then the potential distribution was obtained by solving a direct problem by BEM using the estimated polarization characteristics.

Furthermore, in the estimation of polarization characteristics fuzzy a-priori information was used. By using this result and the above-mentioned optimization method, the optimal current to be impressed from each electrode was determined.

References

[1] Fontana,M.G.& Greene,N.D., *Corrosion Engineering*, Tokyo, Japan: McGraw-Hill,1978.

[2] Strommen,R., Keim,W., Finnegan,J.& Mehdizadeh,P., Advances in offshore cathodic protection modeling using the Boundary Element Method, *Corrosion'86*, Paper 45, (1986), p.45/1.

[3] Adey,R.A., Brebbia,C.A.& Niku,S.M., Application of Boundary Elements In Corrosion Engineering,*in Topics in Boundary Element Research VII*, ed. C.A.Brebbia Berlin, Germany: Springer-Verlag,pp.34-64, 1990.

[4] DeGiorgi,V.G., Kee,A., Lucas,K.E. & Thomas,E.D., Examination of Modeling Assumptions for Impressed Current Cathodic Protection , *Proc. of CORROSION 99 Research Topical Symposium, Cathodic Protection* , USA, NACE International, pp.1-16, 1999.

[5] Riemer D.P. & Orazem,M.E., Models for Cathodic Protection of Multiple Pipelines with Coating Holidays , *Proc. of CORROSION 99 Research Topical Symposium, Cathodic Protection* , USA, NACE International, pp.66-82, 1999.

[6] Deconinck,J., Bortels L., Nelissen,G. & Dam,A., Pynacker,A., Simmulation and Evaluation of a Cathodic Protection System of Buried Pipelinesunder Stray Current Influences , *Proc. of CORROSION 99 Research Topical Symposium, Cathodic Protection*, USA, NACE International, pp.83-89, 1999.

[7] Takazawa,H., Amaya,K. & Aoki,S., Optimum Design of Cathodic Protection for Long Pipelines, *Proc. of Pressure Vessel and Piping 98*, PVP.**370**, New York, USA, ASME, pp.63-69, 1998.

[8] Aoki,S., Amaya.K. & Miyasaka, M., Boundary Elemnet Analysis of Cathodic Protec tion for Complicated Structures, *Proc. of CORROSION 99 Research Topical Symposium, Cathodic Protection: Modeling and Experimaent*, USA, NACE International, pp.45-65, 1999.

[9] Aoki,S., Amaya,K. & Gouka,K., Optimal Cathodic Protection for Ship,*Proc. of Boundary Elemen Technology XI, Southampton, England*, Computational Mechanics Pu blications, pp.345-356, 1996.

[10] Amaya,K., Aoki,S., Takahashi, T., Uragou, M. & Nishikawa, A., B.E. Analysis of G alvanic Corrosion of Long Structure Buried in Soil with Gradient Electrical Conductivity, *Boundary Element 19*, Southampton, England, Computational Mechanics Pub., pp.789-798, 1997.

[11] Aoki,S. & Amaya,K., Boundary Element Analysis of Inverse Problems in Corrosion Engineering, *in Boundary Integral Formulations ʼerse Analysis*, ed. L.C.Wrobel and D.B.Ingham, London, England, (. .putational Mechanics Publications, pp.151-169, 1997.

Experimental vs. computational system analysis

V. G. DeGiorgi & E. A. Hogan
Naval Research Laboratory, Washington, DC, USA

Abstract

Computational and experimental methodologies are often seen as competing resources. This is unfortunate because in the evaluation of complex physical phenomenon often both methodologies are required to develop a complete understanding of the underlying basic science as well as an understanding of the system level behavior. Even though the title implies a contrasting of methods, this work will present how computational methods can be used to supplement experimental techniques and visa versa. Electrochemical corrosion is a complex natural phenomenon that has been studied for many years, however corrosion mechanisms as well as corrosion protection are still not completely defined. The prevention of ship hull corrosion is a major concern for the US Military and the commercial shipping industry due to the large monetary and long-term investment. This paper discusses the physical scale modeling (PSM), shipboard impressed current cathodic protection (ICCP) design technique, how computational methods can be used to augment the design process and presents a number of experimentally determined factors that should be applied to present mathematical code development to improve the computational approach to cathodic protection system design.
Keywords: cathodic protection, boundary element modeling, impressed current cathodic protection, physical scale modeling.

1 Introduction

The need to protect ocean-going vessels from corrosion has existed as long as man has explored the oceans. Today a common approach to corrosion prevention is the use of Impressed Current Cathodic Protection (ICCP) systems. Cathodic protection systems take advantage of the inherent electrochemical characteristics of metals by thermodynamically forcing a reduction (cathodic) reaction on the desired surfaces. The ICCP system thus provides an impressed source of electrons to the metal surfaces, whereby shifting the surface potential from a freely corroding condition to one that favors a cathodic (protected) reaction. Details of the cathodic protection approach can be found in numerous textbooks such as Evans and Morgan [1], [2].

Ships are complex structures that offer challenges to the designers of cathodic protection systems. Not only are material interactions of concern, but operational conditions, surface treatments, biological micro/macro fouling and geometric complexities increase the difficulty of design. Design by rule and design by past experience (empirical designs) have proven not to be sufficient for modern ships. Experimental and computational methods have been proposed independently for ICCP system design. Currently, the accepted design methodology for the U S Navy is physical scale modeling (PSM), which is an experimental approach. However, it is recognized that with proper development and verification, computational analysis can provide significant value.

NRL has been a strong proponent of a combined experimental-computational approach to the design of shipboard ICCP systems. In fact, NRL has applied combined experimental-computational approach to a wide range of research topics. This concept was originally presented for ICCP systems by DeGiorgi et al in 1996 [3]. As advances have been made in both experimental and computational methods, the nature of the interaction between these two approaches has evolved. Originally, as proposed by NRL, experimental and computational studies were independent and sequential. Computational analyses were to be completed first as a preliminary design or screening analysis. Experimental evaluation of a smaller subset of possible ICCP systems would be performed using physical scale modeling techniques. The final design would be selected based on the experimental work.

Today, the combined experimental-computational technique is not proposed as a sequential analysis approach but, rather, as an interactive evaluation approach, that provides direct feedback to both methodologies. This updated approach allows the designer to take full advantage of the strengths of experimental and computational approaches. The present paradigm is a unified experimental-computational approach that requires constant communication between experimentalist and computational analyst. It is a true multidiscipline design approach. Computational methods are no longer seen by the experimentalists as 'just' a design tool but have gained acceptance. This is shown in the recent work where computational methods have been applied to help develop a broader understanding of PSM experimental parameters/allowances such as tank size and geometry, model scale and sensor design [4, 5].

The unified experimental-computational design approach opens up new possibilities in system design. The objective of this work is to briefly introduce the computational community to PSM methods, highlight where the experimental community envisions computational assistance in understanding system performance, and to outline critical areas for combined future work.

2 Physical scale modeling

Physical Scale Modeling (PSM) uses near-exact scale models and scaling factors based on the physics of electrochemical response to provide information on current and potential values on the structure. Kasper [6], Agar and Hoar [7] and Weber [8] define the relationship between scaled geometry, scaled solution conductivity and the resulting interpretation of results. While these early authors addressed theoretical aspects of the problem their work is directed towards electroplating. Theoretical background specifically for ICCP systems and PSM has been presented by Ditchfield et al [9]. Detailed comparisons with real ship data have been used to validate the PSM [10, 11].

The NRL Center for Corrosion Science and Engineering has performed PSM on various ship hulls and hull component geometries since 1987. These models range in scale from full to 1/100th scale and in size from 0.5 to 6 meters in length. A typical representative of ship hull model is shown in Figure 1. The hull is fabricated of fiberglass. Cathodic areas are small-scaled sections of metal attached to the hull as can be seen in the figure. The model is additionally instrumented with scaled ICCP anodes and numerous Silver-Silver Chloride reference cells to provide detail on the hull potential.

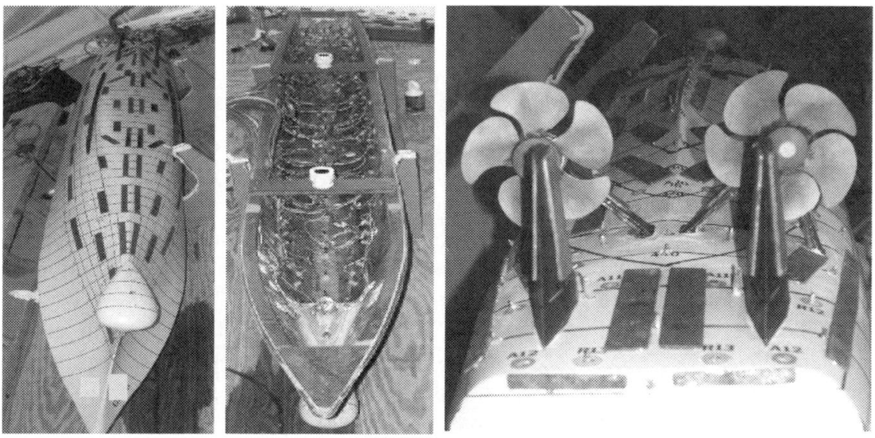

Figure 1: Example of a near-exact scale model used in PSM.

PSM modeling is a robust physics based method for the design of ICCP systems. For both PSM and the computational techniques, cathodic protection behaves in accordance with Ohm's Law:

$$E = I\,(R_P + R_{OHMIC}) \tag{1}$$

where, E = potential (V), I = current (A), R_P = polarization resistance and R_{OHMIC} = electrolyte ohmic resistance. The reader should realize that R_P is highly nonlinear and influenced by environmental conditions. In the scaling required in the PSM technique, $R_{OHMIC} = \rho L/A$, and where, ρ = electrolyte resistivity, L = length of ship or model and A = area. The exact scaling, it is desired for potential relationships to exist, such that $E_{SHIP} = E_{MODEL}$ and for current density (i) behavior, such that $i_{SHIP} = i_{MODEL}$, where $i = A/m^2$ resulting in:

$$E_{SHIP} = I_S(\rho_S L_S/A_S) = i_S(\rho_S L_S) \text{ and } E_{MODEL} = I_M(\rho_M L_M/A_M) = i_M(\rho_M L_M) \quad (2)$$

where it is necessary that $R_{P\,(SHIP)} = R_{P\,(MODEL)}$ for the model to scale exactly, by definition. For scaled models $L_S / L_M = k$ and $\rho_M = \rho_S\,(k)$, where k = scale factor, the relationship becomes:

$$E_{SHIP} = E_{MODEL} = i_S(\rho_S L_M)\,(k) = i_M(\rho_S L_M)\,(k) \quad (3)$$
$$i_S = i_M$$

From (2) and (3) electrical current measured on the model is:

$$I_S = I_M\,(k)^2 \quad (4)$$

There are three basic assumptions in the derivation of the modeling process and that provide for precision measurement of potential and current on the model with a direct mathematical relationship between the model and full scale system. These assumptions are:

- The wetted surface areas and geometry are exact and scaled such that $A_{SHIP} = A_{MODEL}\,(k)^2$.
- The current density relationship, $i_{SHIP} = i_{MODEL}$ is true, by definition, when the model size, electrolyte dilution and polarization resistance components obey the scaling law.
- $R_P = \Delta E/i_C$ must be same for the model and full scale system, where ΔE represents the polarization from E_{CORR} to the cathodic protection set potential of –0.85 V.

The key to the modeling methodology, in addition to the correct implementation of model size and correct electrolyte ohmic scaling is the polarization resistance behavior. In order to preserve the relationship $R_{P\,(SHIP)} = R_{P\,(MODEL)}$ for direct scaling, NRL has established a pre-conditioning sequence. This process allows for the correct cathodic film to be developed on the surfaces of the cathodes so that the R_p response is the same as the full-scale hull. The U.S. Navy has utilized PSM, as performed by NRL Center for Corrosion Science and Engineering, extensively. There is an established design criteria and PSM is currently the accepted standard technique for design of U. S. Navy ICCP systems. This procedure and other details of PSM approach can be found in References [12, 13].

3 Computational methods

PSM can be used to determine ICCP component placement, life-cycle performance, zone interaction behavior and various failure modes under both static (dockside) and dynamic (underway) operational conditions for a variety of ship and system configurations. PSM is able to directly address difficult geometries, areas of restricted flow, protective coatings degradation, advances in ICCP design technology (i.e. use of digital comptrollers – hardware and software, advanced control algorithms) and complex interaction of multiple power supplies and multiple control point locations (i.e. zone behavior).

There are, however, limits to the capabilities of PSM. In terms of shipboard ICCP system design, the limitations are primarily related to the fact that the method uses a scale model representation of the ship hull. It is only possible to fabricate into the model features of a certain dimension. PSM was originally developed for the design of systems that would be operated primarily in open oceans. The scaling of full strength seawater to an appropriate level for testing (as described in the previous section) is possible. However, changes in ship deployment strategy have mandated a closer look at system performance in both brackish and nearly fresh water conditions. It is significantly difficult to scale the conductivity to meet the conditions for testing under these operating scenarios as well as in maintaining an accurate cathodic film due to the extremely low presence or absence of chloride ions in the scaled electrolyte. Finally there is, as always, the cost of testing versus the costs of virtual testing using computational models. Once a hull specific computational model has been created and verified, it is possible to evaluate many scenarios, computationally, without the manpower, time and cost drain of experimental set up and testing. This is true for any experimental methodology.

Computational modeling was seen as a way to supplement the information obtained through experimental techniques. In order to validate computational modeling for ICCP system design and analysis applications, significant work had to be completed to demonstrate capabilities.

4 Priority issues for future development

In reviewing current research the authors have identified areas of concern that should receive priority for future investigation. These are (1) accurate modeling of free flood spaces, including the effects of hybrid systems (combined sacrificial and impressed systems) (2) time-based and spatial variation of polarization design rules, and (3) determining of analysis requirements for on- and off-board electric field calculations. We will briefly discuss each topic.

4.1 Accurate modeling of free flood spaces and hybrid systems

Free flood spaces are confined areas of the underwater hull that have openings to the seawater by means of flood holes, gaps or gratings. Typical of free flood spaces are submarine ballast tanks and surface ship seachests as shown in Figure 2.

As can be seen, the opening and the cavity itself combine to form a complex geometry. These features tend to be small relative to overall underwater hull geometry, however there are variable quantities and sizes on any specific hull, they may or may not have a sacrificial system for local corrosion protection and they incorporated different materials that adds additional galvanic relationships. Experience from ship operations and PSM studies indicate that these features can have an impact on ICCP system performance and the resultant off board electric field signal produced by the electrochemical system [Fig 3].

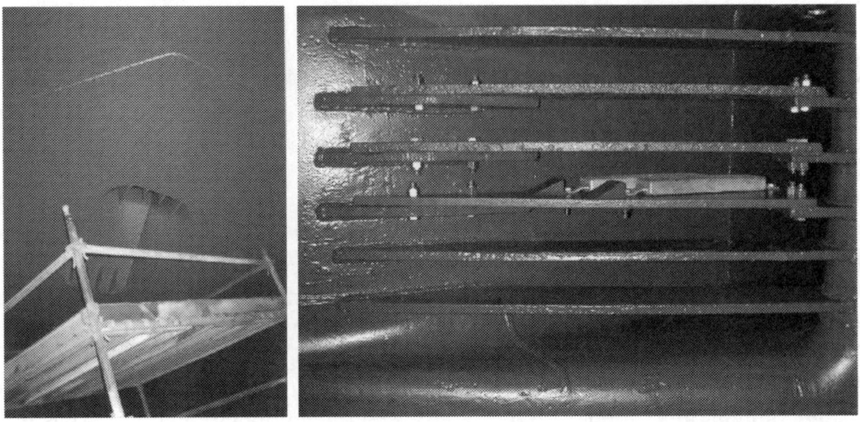

Figure 2: Picture of a Surface Ship seachest opening (left) with splitter bars removed and close-up (Right) with bars and sacrificial anode installed.

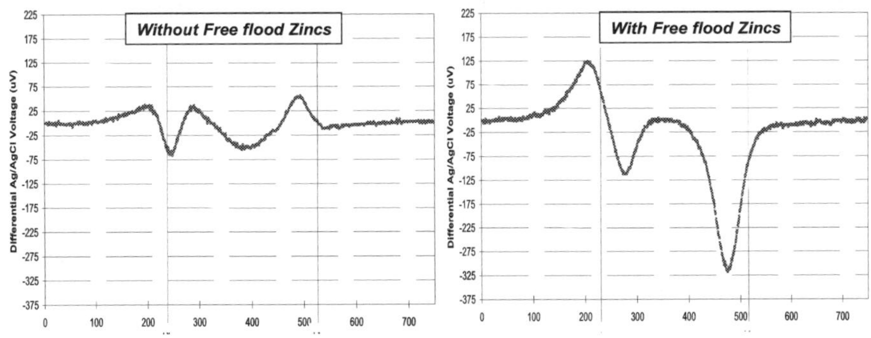

Figure 3: Influence of zinc anodes in free flood spaces on the off-board electric field signal measured during PSM.

There are several technical issues that must be addressed involving these features. It is impractical to incorporate all free flood spaces in PSM test articles. However, it should be possible to incorporate all of these features in computational models. Should these features be included directly in a hull model or evaluated through use of submodeling, must be investigated. Of concern is the applicability of the LaPlace solution approach, used in determining electrochemical corrosion related current and voltage values, in the incorporation of these features in a computational model. The LaPlace solution requires that the electrolyte is of uniform composition. Free flood spaces are small volumes with stagnant regions. Without the free flow of electrolyte along all surfaces, there will be variations in oxygen and other seawater chemical components resulting in different and varying material response characteristics for the subspace environment. It is possible that LaPlace solution is no longer valid. This is a major concern.

4.2 Spatial and time-based variation of polarization design rules

It is known that a key to accurate computational modeling is the use of an appropriate polarization curve to describe the material response. It is also known that polarization response is a sensitive nonlinear material parameter that has been shown to be affected by numerous environmental factors. Recent work has demonstrated that there is a spatial variation of polarization response along the hull of a surface ship as shown in Figure 4. These polarization curves were determined by direct measurement of polarization response at different points along the hull of a PSM test article. A single curve can be generated as shown in by the curve fitted through the squares in Figure 4, but the use of only this curve would greatly limit the accuracy of the final calculated hull potentials and currents.

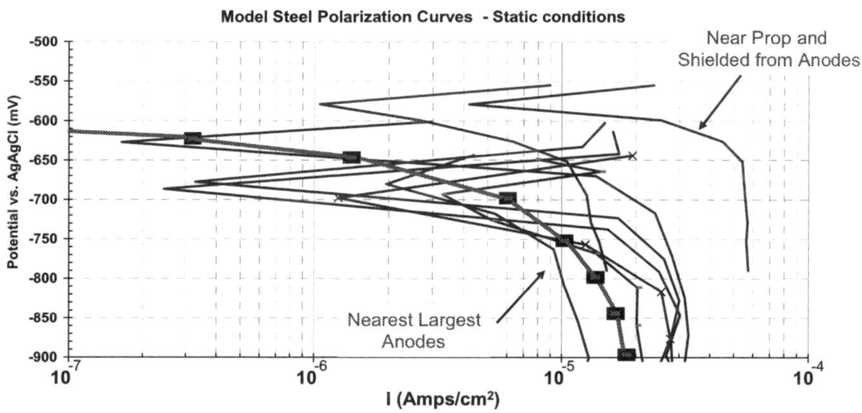

Figure 4: Polarization material response differences as a function of spatial relationship to anodes and noble metals.

The variability in the curves can be related to a time-based phenomenon, whereby hull metal that is near anodes and is not shielded will develop the natural cathodic film faster relative to other more shielded regions on the hull. This calcareous film helps to reduce the cathodic demand of the cathode which results in a shift in the material curves to a lower current density demand. It is possible to assign element regions within a computational model with the appropriate polarization response based on this type of measured data. The mapping of measured values to a computational model implies that there has to be a one to one correspondence between ICCP system on the test article and ICCP system as defined on the computational model. This eliminates a major advantage of using computational methods to evaluate design alternatives.

What is needed is a series of design rules that automatically assigns appropriate polarization response to a region on a computational model based on anode location, cathode location, and material type as well as incorporates a time-based correction to account for calcareous film formation. This would allow the designer to place anodes and cathodes as required with the knowledge that their region of influence will be incorporated into the computational model. The design rules must be based on understanding of the range of influence features such as anodes and noble metals have on the 'basic' polarization response'

4.3 Analysis requirements for on- and off-board electric field calculations

The analysis of real ship behavior is a complex relationship of many of the factors discussed, where these data are taken in real time, using the actual geometry, material relationships and parameters inherent of a natural system. In a simulation of these behaviors, the accuracy of the field relationships measured depend on precision, correlation, magnitude and spatial equivalency of the field behavior and computationally, these intrinsic properties generally come from experimentally determined values feeding the code. The results of previous boundary level models, limited by computer memory and space, produced geometrically simple models where components, such as the propellers, were stylized to provide for the same cross sectional boundaries without the need for the complex detail, illustrated in Figure 5. This subsequent detailed investigation has shown that the geometric detail is necessary and that incomplete geometric description provides significant error in simulation. Error can also be cumulative in nature because near-field effects directly impact the nature of the off-board integration. Errors attributable to localized current density distribution, geometric shielding and variation in materials polarization properties directly impact the primary near-field computational relationships and magnify the error of calculated far-field relationships.

PSM treats the hull as if it were a real ship, except that there are also limitations in what can be modeled and represented. Experience has shown that the computational equivalent has many benefits when expressing and considering numerous variables, but the significance of the data and verification must be fully tested as a design instrument, not just compared against known data.

In addition to the importance and type of the data input into the code, it is the code itself that needs critical review. In most cases, the LaPlace equation that describes the fundamental behavior of the ICCP system, but these relationships are also conditional on certain equilibrium states. The review of data, such as that for tanks, shows a diffusion effect and polarization behavior that falls outside of some computational aspects and may even result in a non-invertible matrix situation. Thus, the code requires the addition of mathematics to define areas with unique boundary properties, diffusion and current crowding. Additionally, structures often have multiple zones in single ICCP systems and have multiple ICCP systems, giving rise to “zone" behavior. Each zone can interact and computationally, it is not known quite how these behaviors relate or how an actual system response compares in performance. This situation can be further complicated by the existence of multiple ICCP systems on a single structure.

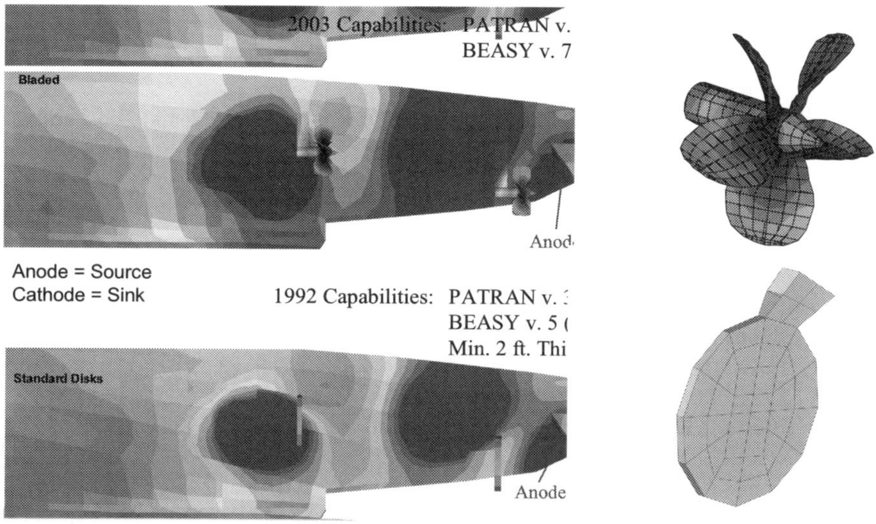

Figure 5: Effect of Geometric mesh detail on computationally derived surface potentials.

5 Summary

Design of an effective ICCP system is critical to maintain the Navy’s long-term investment and fleet readiness. Presently, a combined experimental and computational approach to design has proven to be the most comprehensive. Further understanding of the electrochemical response of an ICCP over the ships lifecycle and environments is needed to shift the combined approach to be more dependent on the computational results. Ideally this will lead to the situation where design will be totally based on computational analysis. This paper has discussed three major areas where further understanding is required and needs to be addressed in future computational programs.

Acknowledgements

The support of Dr. A. I. Kaznoff and Mr. E. D. Thomas, Naval Sea Systems Command, is gratefully acknowledged. The authors also wish to thank Mr. K. E. Lucas, NRL Code 6130, and Dr. S. A. Wimmer, NRL Code 6350, for their assistance.

References

[1] Evans U. R., Metallic Corrosion Passivity and Protection, Edward Arnold and Co, 1948.

[2] Morgan J., Cathodic Protection, NACE, 1987.

[3] DeGiorgi, V. G., Thomas, E. D., Lucas, K. E. & Kee, A., Boundary Element Technology XI, 335-345, 1996.

[4] Wimmer, S. A., DeGiorgi, V. G. and Hogan, E., Boundary Element Technology XV, 213-222, 2003.

[5] DeGiorgi, V. G., Wimmer, S. A., Hogan, E. & Lucas, K. E., Boundary Element Method Vol. XXII, 439-447, 2002.

[6] Kasper, C., Trans. of Electrochemical Society, Vol. 77 and 78, Papers I-IV, 1940.

[7] Agar, J. N. & Hoar, T. P., Discussions of Faraday Society, No. 1, 162-168, 1947.

[8] Weber, J. T. & Fagen, B., J. of Electrochemical Soc., Vol. 103, No. 1, 64-72, 1956.

[9] Ditchfield, R. W., McGrath, J. N. & Tighe-Ford, D. J., J. of Applied Electrochemistry, 25, 54-60, 1995.

[10] McGrath, J. N., Tighe-Ford, D. J. & Hodgkiss, L., Corrosion Prevention and Control, April, 33-36, 1985.

[11] Thomas, E. D. & Parks, A. R., Proceedings Corrosion 89, Paper No. 274, NACE, 1989.

[12] Lucas, K.E., Thomas, E. D., Kaznoff, A. I. & Hogan, E. A., Designing CP Systems for Marine Structures and Vehicles, STP-1370, ASTM, 17-33, 1999.

[13] DeGiorgi, V. G., Hogan, E., Lucas, K. E. & Wimmer, S. A., chapter in BEM and Corrosion, WIT Press, In Press.

Numerical 3D BEM simulation of a CP system for a buried tank influenced by a steel reinforced concrete foundation

M. Purcar[1], J. Deconinck[1], B. Van den Bossche[2], L. Bortels[2] & P. J. Stehouwer[3]
[1]*Vrije Universiteit Brussel, Dept. of Electrical Engineering, Belgium*
[2]*ElSyCa N.V., Zellik, Belgium*
[3]*Gasunie N.V., Groningen, The Netherlands*

Abstract

An advanced three-dimensional Boundary Element Method (BEM) approach is used to quantify the influence and mitigation of a steel reinforced concrete foundation on the cathodic protection (CP) level of a neighbouring buried tank. Due to the fact that the reinforced concrete is connected to the CP system, this concrete structure attracts a substantial fraction of the current that is injected by far field anodes. As a consequence, the tank wall that is facing the foundation, is under protected (tank/soil potential *-600* to *–700 mV* vs. Cu/$CuSO_4$), while the opposite tank wall (facing the far field) is well protected (*-1200 mV* vs. Cu/$CuSO_4$). These simulations are in agreement with the measurements. Rather than remediate this situation by an expensive trial and error approach, Gasunie preferred to cooperate with ElSyCa and to simulate first the effects of various possible solutions they had in mind. These simulations predict a significant improvement of the tank wall protection level by about *200* mV at the foundation side when insulating screens are placed between the tank and the foundation (depending on size and position of the screens) while the far field side is hardly influenced. As the solver is fully integrated in the SolidWorks CAD environment, any modification of the dimensions of the entire configuration (e.g. the screens), the surface mesh characteristics (determining the computational accuracy), and the physico-chemical input parameters (polarisation characteristics, soil conductivity, etc.) can be established in a fast and easy manner.

Keywords: cathodic protection systems, advanced software packages, advanced BEM.

1 Introduction

Cathodic Protection (CP) systems are widely applied to buried tank and pipeline structures, as they compensate (by electrochemical means) for the loss in physical protection due to the degradation of tank coatings over time. Most often, these CP systems contain a series of impressed current anode ground beds (though also sacrificial anodes can be used), placed at a remote distance from the pipes and tanks. The entire configuration of the CP system and the buried structures has some particular characteristics that necessitate and justify the use of numerical simulations. First, the hidden character and low accessibility of these structures make installation, maintenance and repair very expensive. Also, the geometry of most steel structures that are subject to cathodic protection is too complex to allow analytical or even empirical estimations for the determination of the local protection level. Numerical modeling provides significant relieve by pointing out insufficiently protected regions - possibly subject to corrosion, and overprotected regions - subject to hydrogen evolution and hence coating delamination or hydrogen embrittlement. As a consequence, numerical modeling allows simplification and optimization of installation, maintenance and repair. Moreover, models also provide reference values for measurements on operational sites, enabling to trace and solve any possible anomaly.

The Boundary Element Method is a well suited for the computation of CP system performances on buried or submerged structures. Orazem et al. [1,2] use a 3D BEM approach to compute the protection level of large coating defects on pipelines. Results are presented for a pipe segment of limited length (10 feet), in presence of a parallel anode system. Riemer and Orazem [3] also produced results for a larger pipeline (> 6 km) with coating defects of varying size, and investigated the ability of coupons in the vicinity of the defects to measure off-potentials. Adey applied a full 3D approach in order to allow computation of the potential field in the neighbourhood of jacket joints under cathodic protection of sacrificial anodes [4]. The present authors used a 3D coupled multi-domain BEM approach to simulate the protection level of a buried pipe segment surrounded by a U-shaped concrete vault [5].

This paper presents results for an advanced 3D BEM approach, applied to quantify the influence and mitigation of a steel reinforced concrete foundation on the cathodic protection (CP) level of a neighbouring buried tank. Also the insertion of a vertical insulating screen reducing the effect of the concrete foundation on the CP level of the tank is simulated and investigated.

2 Mathematical model and solution method

The most important electrochemical reactions that may take place on a bare steel surface are, depending on the local pipe-to-soil potential: metal oxidation, reduction of dissolved oxygen and hydrogen evolution. For CP-simulations the steel electrode polarisation is often encompassed in one single, measured electrode polarisation curve [6], relative to a mixed corrosion potential E_{corr}:

$$j_n = f(V - U - E_{corr}) = f(\eta) \tag{1}$$

with U the potential in the soil adjacent to the electrode, V the metal potential and $\eta(j_n)$ the polarization overvoltage being a function of the local current density j_n. For the simulations in this paper, the corrosion potential E_{corr1} (= *-476 mV* vs Cu/Cu-sulphate) and the polarisation behaviour are taken as in reference [6], for soil type B (sand).

2.1 Coated steel surface

As the tank is covered with a very good quality coating, it is assumed that the latter represents a resistivity that is in series with the pipe / soil polarisation resistance:

$$j_n = f(V - U - E_{corr1} + \rho L_{coat} \cdot j_n) \tag{2}$$

where the value for ρL_{coat} is $8.0E{+}4\ \Omega.\ m^2$, yielding a current density $j = -5.0E\text{-}6\ A/m^2$ for a protection level $V\text{-}U$ of $-850\ mV$ vs Cu/Cu-sulphate. Due to the high resistivity of the coating, curve (2) manifests a nearly linear behavior.

2.2 Concrete surface

For the concrete foundation, the concrete acts as an additional resistive layer between the steel skeleton and the soil. The corrosion potential steel / concrete takes an entirely different value $E_{corr2} = -\ 200\ mV$ vs Cu/Cu-sulphate (compared to E_{corr1}). Neglecting the steel / concrete polarisation, and assuming a linear behaviour for the concrete resistance allows to describe the steel / concrete / soil interface by:

$$-U - E_{corr2} = \rho_{concrete} \cdot L \cdot j_n \tag{3}$$

where L holds for the mean thickness of the concrete between the steel skeleton and the soil (e.g. *5 cm*), and $\rho_{\text{concrete}} = 3000\ \Omega.m$ is a typical concrete resistivity value.

2.3 Soil medium

Tanks buried in the soil are vast electrodes immersed in a semi-infinite electrolyte. Due to the nature of the conductive medium and its large scale, no concentration gradients occur in the electrochemical solution except for the well-known and very thin diffusion layer near the electrodes. Since the soil is assumed to feature only charge transport with normal ohmic resistivity effects, the potential model holds, being described by the Laplace equation:

$$\overline{\nabla}(\overline{j_n}) = 0 \qquad\qquad \overline{j_n} = -\sigma\overline{\nabla}U \tag{4}$$

The phenomena occurring in the diffusion layer and at the electrode interface are encompassed in the (non-linear) boundary conditions as described by equation (1).

For domains with a constant electrical conductivity σ, equation (4) simplifies to:

$$\overline{\nabla}(\overline{\nabla}U) = 0 \tag{5}$$

On insulation boundaries, the current density perpendicular to the surface should be zero, which results in the following boundary condition:

$$\overline{j_n} \cdot \overline{1}_n = j_n = -\sigma \overline{\nabla} U \cdot \overline{1}_n = \sigma\, Q = 0 \tag{6}$$

The soil conductivity for the simulations in this paper is taken *4.8E-3 S/m* (wet sand).

2.4 Numerical solution method

As the conductivity of the computational domain is constant, BEM is to be preferred over other discretisation methods as Finite Element (FEM), Finite Volume (FVM) and Finite Difference Methods (FDM), to solve the simplified charge conservation equation (5). When the BEM is applied, only the boundaries of the domain are to be discretised. Another advantage, particularly for electrochemical systems, is that the current density distribution along the electrodes is a direct unknown to this problem, rather than a variable that has to be computed afterwards from the derivative of the potential field perpendicularly to the electrodes. This turns BEM into a more accurate method for potential problems, compared to FEM, FVM and FDM.

The characteristic BEM equation for the contribution to node *i* is [7,8]:

$$c^i U^i + \int_{\Gamma} U \frac{\partial w^*}{\partial n} d\Gamma = \int_{\Gamma} w^* Q \, d\Gamma \tag{7}$$

with $w^* = \dfrac{1}{4\pi r}$ the 3D Green function and $Q = \dfrac{\partial U}{\partial n}$ the flux at the boundary nodes. c^i is for the space angle in node i. Γ is the 2D surface that encloses the 3D computational domain.

In order to apply BEM, the boundary Γ is to be discretised into a series of N non-overlapping elements, transforming equation (7) into:

$$c^i U^i + \sum_{j=1}^{N} \int_{\Gamma_j} U \frac{\partial w^*}{\partial n} d\Gamma = \sum_{j=1}^{N} \int_{\Gamma_j} w^* Q \, d\Gamma \tag{8}$$

The index j ranges over all elements of the domain and integration is performed over the surface Γ_j of each element.

Triangular elements with linear shape functions for the unknown potential field are used for the 3D BEM computations as presented in this paper, restricting the unknowns to the nodal values, N in total.

Taking into account equation (6), the BEM system of equations can be expressed in matrix form:

$$[H] \cdot \{U\} = [G] \cdot \{J_n\} \tag{9}$$

with $\{U\}$ and $\{J_n\}$ unknown vectors of size N. When the boundary conditions have been introduced, i.e. zero current on insulators and overpotential relations (2) and (3) on electrodes, one yields:

$$[H] \cdot \{U\} = [G] \cdot \begin{Bmatrix} J_n = f(V - U) \\ J_n = 0 \end{Bmatrix} \tag{10}$$

The matrices H and G in equation (10) are fully populated and depend only on the geometry of the system and the conductivity of the domain. This system of equations can be rewritten to group all unknowns in the left hand side vector:

$$[H^*] \cdot \begin{Bmatrix} J_n \\ U \end{Bmatrix} = [G^*] \cdot \begin{Bmatrix} V - \eta(j_n) \\ 0 \end{Bmatrix} \tag{11}$$

In most practical cases, system (11) is non-linear, except when $\eta(j_n)$ is linear in j_n.

The non-linear system equation matrix is solved using a Newton-Raphson iterative method [7], combined with a standard Gauss algorithm to tackle the linear system of equations that appears during each iteration of the Newton-Raphson procedure.

The electrode polarization data can be specified either as a linear or an exponential (Butler-Volmer) expression. Also a numerical data set can be used, in this case the data couples $(V\text{-}U, j_n)$ are transferred into a continuous derivable curve using a spline polynomial function.

A hybrid grid generator is used to produce the surface mesh that is required for the BEM computations [10]. This flexible grid generator creates an unstructured or structured mesh for each surface separately, or an unstructured mesh with structured boundary zones. The construction of a mesh that is optimally refined towards zones where a steep variation of field variables is expected, and that generates coarse elements at any other position, allows to produce reliable numerical results with a restricted number of elements, hence limiting the computational efforts to solve the system of equations to a minimum.

3 Application

3.1 Configuration

The configuration of interest consists of a buried tank with concrete foundation (the latter assumed non-conducting) and a part of a neighbouring steel reinforced concrete foundation of a building (figure 1). The foundation is electrically connected to the CP system, hence exhibits a negative impact on the protection level of the tank surface. An insulating screen is installed between the tank and the foundation in order to limit the impact of the concrete foundation. The tank has a diameter of *1.3 m*, *4 m* length and is positioned about *0.9 m* below the earth surface. For computational purposes, a sufficiently large box has been defined that surrounds the configuration of interest, the top plane coinciding with the earth surface and the side wall opposite to the tank being identified with the far field. Also a series of small cubes has been defined, representing possible positions of the reference electrode in perforated probe tubes. These small cubes are considered insulating, hence do not affect the current density distribution in the soil, they simply take the potential of the surrounding soil.

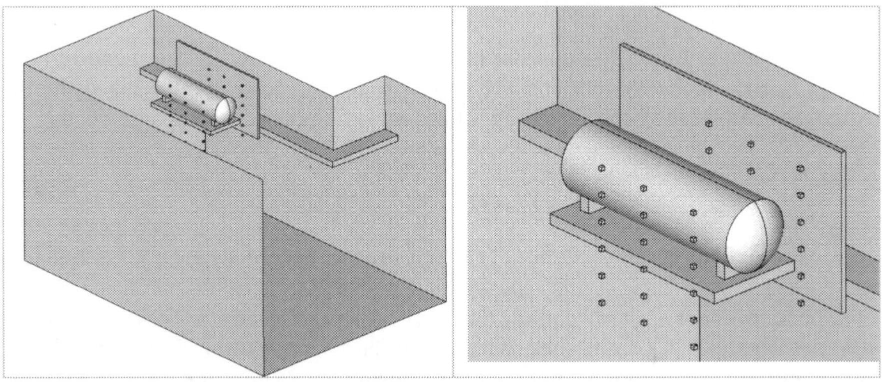

Figure 1: Configuration of tank + foundation + screen.

3.2 Boundary conditions

A far field potential = *1.75 V* is imposed at one face of the surrounding box. The metal potential of both the tank metal and the concrete foundation is set at *0 V*. All remaining surfaces are insulating.

3.3 Surface mesh

A fine triangular mesh is used for the tank surface, while for example the faces of the surrounding box are discretised with a coarse mesh. This ensures an optimal trade off between computational efforts and accuracy of the simulated results.

3.4 Computational specifications

The entire configuration was defined in the SolidWorks© CAD package, and simulations were performed with the CP-Master© v1.0 software tool, which is entirely integrated in this CAD package [11].

The total number of elements used to solve the BEM problem was about 44 000 for the configuration with screen, and 8 000 without the screen.

The CPU time is about *5 hours* on a *2 GHz* Pentium IV PC for 4 iterations in the Newton-Rapson procedure (configuration with screen), after which a sufficient convergence level was reached. For the configuration without screen, the CPU time is below *15 minutes*.

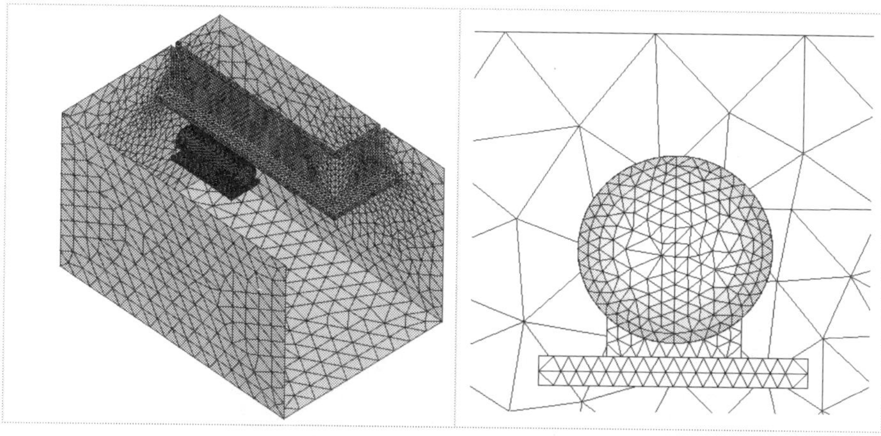

Figure 2: Triangular surface mesh for the entire configuration without screen (left) and detail of the tank surface (right).

4 Results and discussion

The soil potential U, adjacent to all surfaces of the configuration, is obtained as a direct result from the BEM computations (figure 3). The local pipe-to-soil potential (vs the $Cu/CuSO_4$ reference electrode) is directly available from this variable field:

$$V - U = 0.0 - U \tag{12}$$

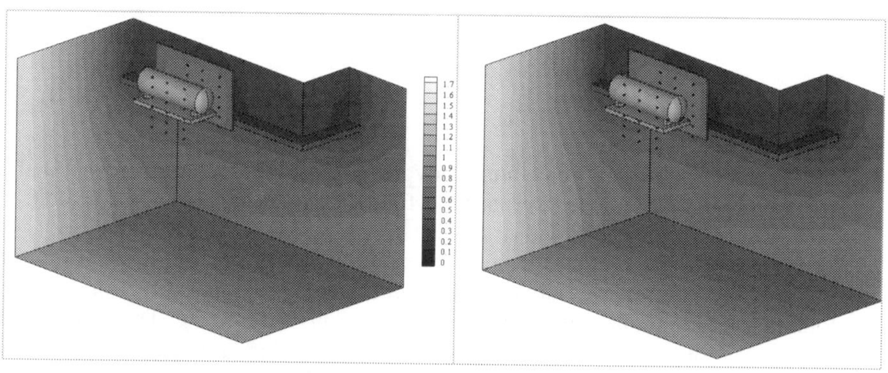

Figure 3: Soil potential distribution over all surfaces.

The simulations for the configuration without screen predict a protection level of about *–1100 mV* vs $Cu/CuSO_4$ at the far field side of the tank, and *–700 mV* at the foundation side (soil potentials observed directly at the tank surface). It is also noticed that the position of the reference electrode in the probe tubes generates a small systematic measuring error, since there is an additional potential drop of around *100 mV* between the reference electrode positions (represented by the small cubes) and the tank surface (at both sides). Taking into account this systematic measuring error, the protection level of the tank surface is respectively *–1100 mV* and *–700 mV*, very close to the values as measured in practice. A detailed view on the protection level over the tank surface is given in figure 4.

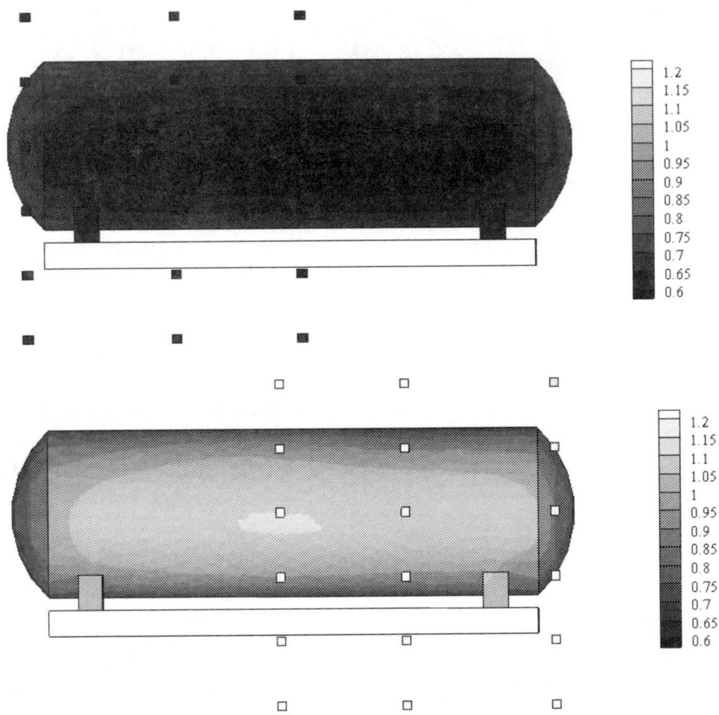

Figure 4: Soil potential distribution over the tank surface (no screen) at the foundation side (top) and far field side (bottom).

Upon inserting the screen, the simulations predict a protection level at the far field side of the tank that remains nearly unchanged, while the protection level at the foundation side mounts with about *150* to *200 mV* (figure 5). This brings the lowest protection levels to about *–800 mV* vs $Cu/CuSO_4$ (including the systematic error induced by the probe tube positions), whereas *–850 mV* is often considered a minimum level for good practice in CP systems.

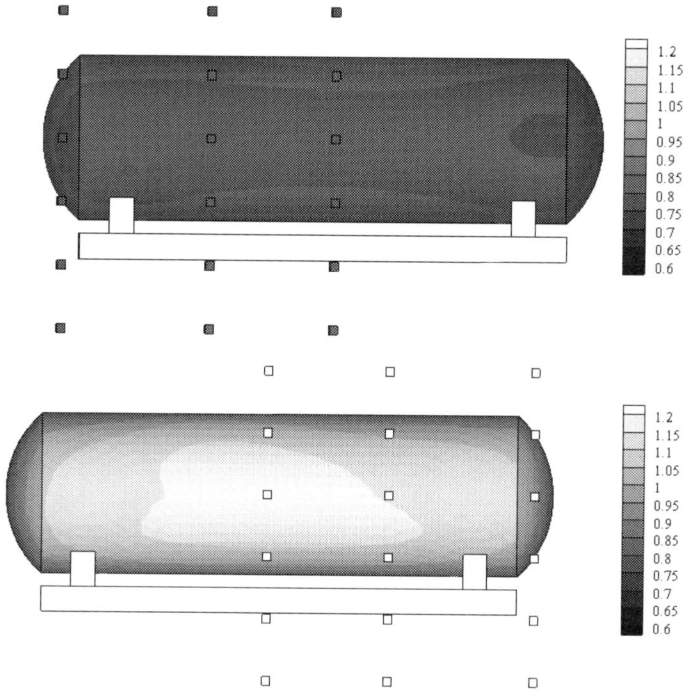

Figure 5: Soil potential distribution over the tank surface (with screen) at the foundation side (top) and far field side (bottom).

5 Conclusion

The simulations predict a significant influence of the screen on the tank surface protection level, but the minimum protection level of *–850 mV* vs Cu/$CuSO_4$, a value generally accepted for good practice in CP system engineering, is not attained at any position on the surface.

Due to the fact that the entire configuration is defined in a professional CAD environment, further optimisation of the screen can be performed in a fast and flexible manner.

References

[1] M.E. Orazem, J.M. Esteban, K.J. Kennelley, R.M. Degerstedt, Corrosion 53(1997)427.

[2] M.E. Orazem, J.M. Esteban, K.J. Kennelley, R.M. Degerstedt, Corrosion 53(1997)264.

[3] D.P. Riemer, and M.E. Orazem, Corrosion 56(2000)794.

[4] R.A. Adey, Topics in Boundary Element Research, vol. 7, Electrical Engineering Applications, chapter 3, Springer-Verlag Berlin, Heidelberg 1990.

[5] M. Purcar, B. Van den Bossche' L. Bortels' J. Deconinck, P.Wesselius, Corrosion 59 (2003) 1019.

[6] D. Rabiot, F. Dalard, J.J. Rameau, J.P. Caire, S. Boyer, J. Applied Electrochem. 29(1999)541.

[7] Deconinck J., Current distributions and Electrode Shape Changes in Electrochemical Systems, Lecture Notes in Engineering 75, Springer Verlag Berlin, ISBN 3-540-55104-2, 1992.

[8] C.A. Brebbia, Boundary Element Techniques – Theory and Applications in Engineering, Springer-Verlag Berlin, Heidelberg, 1983.

[9] J. Deconinck, G. Maggetto, J. Vereecken, J. Electrochem. Soc., 132 (1985) 2960.

[10] Athanasiadis A.N., Deconinck H., Int. J. Numer. Meth. Eng. 58 (2003) 301.

[11] CP-Master v1.0, User Manual, www.elsyca.com.

Section 2
Modelling methodologies

Galvanic corrosion modeling in flowing systems

D. A. Shifler
102 Potomac Street, Boonsboro, MD 21713-1211, USA

Abstract

Contact between dissimilar alloys will generate galvanic currents based on a variety of factors. The extent of risk to the overall systems and specifically to the more electronegative alloy materials can be clarified if the magnitude of the corrosion rate at various locations can be modeled and calculated.

When a more electropositive cathode alloy is separated from the electronegative anode alloy in a complex geometry, there will be a variable distance between the anode and cathode and fluid flow characteristics may influence variations in the limiting current density. In systems containing elbows, the current and potential distribution will not be only in the longitudinal direction, but will have a radial contribution to current flow and corresponding potential changes.

Keywords: galvanic corrosion, current distribution, flow, modeling, piping, boundary layer, polarization, dissimilar alloys, bimetallic corrosion.

1 Introduction

The effect of coupling two different metals/alloys together, either directly or through an external path, increases the corrosion rate of the anodic alloy (the material with the more electronegative potential) and reduces or suppresses the corrosion rate the cathodic alloy (the material possessing the more electropositive potential). The electrochemical degradation derived from the joining of two or more different or dissimilar alloys is termed galvanic corrosion.

Galvanic corrosion causes failures in a wide variety of environments such as atmospheric, automotive, seawater, soil, freshwater, or oil and gas.

The necessary conditions for galvanic corrosion require: (1) dissimilar alloys, (2) electrical contact, either directly or by a secondary, grounding path, between the dissimilar alloys, and (3) an electrolyte. The coupling of dissimilar alloys in conducting, corrosive solutions such as seawater, oil and gas, and biological fluids can lead to accelerated corrosion of the more anodic, electronegative alloy and protection of the more cathodic, electropositive alloy. The extent of galvanic corrosion depends on factors such as: (1) the type of joint; (2) the effective area ratio of the anodic and cathodic members of the galvanic couple; (3) geometry of the coupling members; (4) mass transport (convection, migration, and/or diffusion); (5) bulk solution properties (oxygen content, pH, conductivity, pollutant/contamination level); (6) temperature; (7) flow rate; (8) volume of solution; (9) major and minor alloy constituents; (10) protective film characteristics (oxide, salt films, etc.) in the given environment; (11) surface condition; (12) reaction kinetics (metal dissolution of the anodic member, and either oxygen reduction or hydrogen evolution overvoltages, or the cathodic efficiency of the more noble, cathodic alloy); and (13) the difference in potential between the alloys [1, 2]. Figure 1 shows these and other factors that affect galvanic corrosion [1].

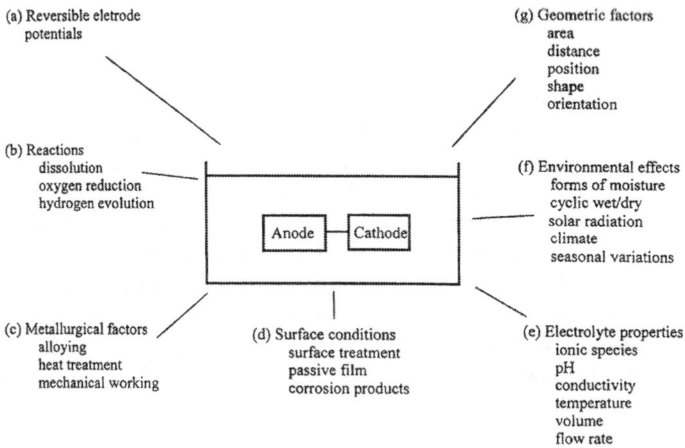

Figure 1: Factors affecting galvanic corrosion [3].

Usually galvanic couples involve electrolytes containing dissolved oxygen. When the galvanic current is limited by oxygen diffusion to the cathode, it is under cathodic control. Under this condition the galvanic current rate is directly proportional to the cathode area and independent of the anode area [3]. Later referred to as the catchment area principle, it also illustrates that the anodic current density is inversely proportional to the anode area. Therefore, the area of the cathode should be minimized so that the cathode/anode area ratio is very small for better long-term performance.

The composition of the electrolyte influences the magnitude, the distribution, and often the direction of the galvanic current flow. High-conductivity fluids tend to spread out the galvanic currents over larger surface areas, when possible, while low conductivity fluids will limit the galvanic current distribution to just the cathode/anode junction [4].

Galvanic corrosion is sensitive to soluble ions in the electrolyte that affect the stability of passive films on the cathodic member of the galvanic couple or the kinetics of the corroding anodic member. The presence of anions such as bicarbonate, silicate, and sulfate may promote the development of insoluble, adherent corrosion products that protect the anodic member from further corrosion and change the relative electrochemical potentials of the galvanic couple based on the relative oxygen concentrations at the surface of each alloy. Other organic or inorganic complexing ions may accelerate corrosion of particular metals and alloys.

The magnitude of the potential difference between the dissimilar materials does not indicate the degree of galvanic corrosion because potentials are a function of the thermodynamics and not of the reaction kinetics that may occur. A difference of 50 mV between select dissimilar materials can lead to severe corrosion while a potential difference of 800 mV has been successfully coupled [5].

Temperature influences the magnitude and direction of galvanic corrosion current. For a zinc/steel or aluminum/steel couple, increasing temperature above 60°C or more may result in polarity reversal due to a change primarily in the zinc potential aided by the presence of bicarbonates and nitrates in the water [6,7].

When dissimilar alloys are coupled, the rate of dissolution at the anode or the reduction reaction rate at the cathode is not equal for all metals/alloys. The reaction rate at the anode or cathode surface determines the electrode efficiency. Polarization caused by current flow between the dissimilar metals/alloys causes a shift in the electrochemical potential of the anodic alloy to a more electropositive value and a shift of the cathodic member to a more electronegative value. The extent of polarization will depend on the metal/alloy and the specific environment to which the galvanic couple is exposed. Generally, in neutral environments, the galvanic couple will be under cathodic control. The extent of cathodic polarization will determine how well a material will drive the corrosion of the anode. Highly polarizable cathodes (alloys with low cathodic efficiency) will tend to cause relatively benign galvanic corrosion, while high cathodic efficient materials can promote severe galvanic corrosion. However, cathodic efficiency can change under different conditions that affect the material surface kinetics [5]. Galvanic relationships between different alloys may be altered in sulfide-polluted seawater [9] because of changes in the stability of the passive films.

Using computer models to predict the magnitude of galvanic corrosion or the current demand for cathodic protection systems in seawater or other electrolytes requires accurate polarization data for the materials involved [9]. However, polarization curves for most structural materials are exposure-time and scan-rate dependent, thus complicating long-term galvanic corrosion predictions [9].

Platinum, silver, and copper form naturally thin formed oxide films and are easily reduced to metal; these alloys are efficient cathodes and, therefore, tend to promote galvanic corrosion [5]. The increasing use of nonmetallic materials (metal-reinforced composites, graphite metal matrix composites, conducing polymers, semi-conducting metal oxides, and conducting inorganic compounds) often exhibit potentials that are more electropositive than most alloys, and when coupled to most alloys, cause galvanic corrosion [10].

In contrast, aluminum, stainless steels, and titanium all have a stable oxide film and tend to polarize and hence, are poor cathodes. In flowing, aerated seawater the oxide film is likely to thicken, thus diminishing galvanic corrosion of the coupled metal further.

2 Fluid flow

Fluid movement, usually characterized by velocity or dimensionless Reynolds number, can have an extreme effect on the corrosion behavior of metals and alloys in fluids. An increase in fluid flow may increase corrosion rates by removing protective films or by increasing the diffusion or migration of deleterious species. Conversely, increased flow may decrease corrosion rates by eliminating aggressive ion concentration or enhancing passivation or inhibition by transporting the protective species to the fluid/metal interface. Thus flow, in combination of other factors, may strongly influence galvanic corrosion.

In any given velocity domain, local velocities permutations may exist over diverse areas of a component due to factors such as geometry, flow disruptions, or mode of fabrication [11]. Turbulent flow increases agitation of waters with the structural materials more than in laminar flow, with the resultant probability of increased galvanic corrosion.

Flow above a critical velocity in some alloys such as copper-based alloys are susceptible to a critical surface shear stress in seawater where the corrosion product film begins to breakdown and is removed from the alloy surface, subsequently causing accelerated attack [12].

Titanium and some nickel-chromium-molybdenum alloys perform well in low, intermediate, and high velocity ranges up to 120 fps (36 m/s).

3 Galvanic corrosion: effects by the environment

The effective galvanic length in piping systems is the distance that galvanic current flows due to the presence of a dissimilar metal couple and may range from 15 to over 200 pipe diameters [13-16]. Steel coupled to stainless steels and copper-nickel will have the tendency to form calcareous deposits on the cathodic members of the couple, thereby reducing galvanic currents.

The galvanic corrosion rate of many copper and nickel-based alloys, and of stainless steels in seawater, depends upon the flow rate as well as the area ratio. Copper and copper-nickel alloys tend to become more noble, i.e., more electropositive, and corrode less as the flow rate increases. In well-aerated, flowing solutions nickel-based alloys and stainless steels are also likely to become more passive and corrode less than under stagnant conditions [5].

In general, the rate of corrosion of coupled and non-coupled metals decreases with exposure time. This is generally due to the diminishing rate of diffusion of oxygen (or hydrogen) through the corrosion product films at cathodic regions and corresponding protection afforded by the corrosion product at anodic sites.

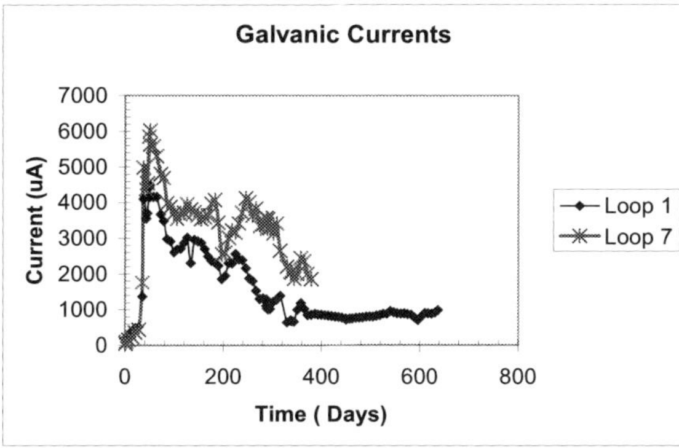

Figure 2: Titanium/70:30 Copper-nickel piping couples-measured galvanic currents per exposure time.

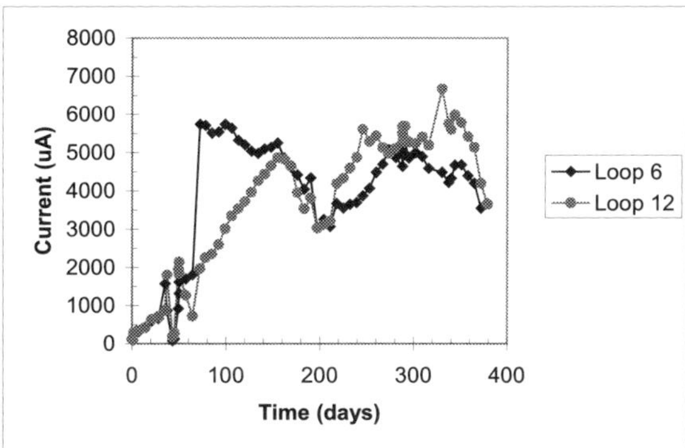

Figure 3: Alloy 625/70:30 Copper-nickel piping couples-measured galvanic currents as a function of exposure time.

Research was done to observe the galvanic corrosion effects of coupling straight titanium alloy or Alloy 625 piping sections to straight 70:30 copper-nickel piping in flowing (at 6 fps (1.8 m/s)) seawater [17]. Both titanium and Alloy 625 pipes had similar cathodic potentials and generated similar galvanic currents when coupled to copper nickel piping. Both uncoated titanium and Alloy 625 piping sections experienced ennoblement of several hundred millivolts between 14 and 65 days. The galvanic currents increased 7-14 days after ennoblement occurred as shown in Figures 2 and 3. Ennoblement of the cathode (titanium in this case) increased the potential difference between the anode and the cathode as shown in Figure 4. The phenomenon of ennoblement is likely caused by microbial action that increases the cathodic kinetics, which shift the cathodic polarization curves to higher currents, thus increasing galvanic currents on the anode [18-20]. It is also observed during the 600+ day exposure that the galvanic currents and the potentials of both the cathode and the anode varied. The cathodic potential and the resulting galvanic current tended to decrease with longer exposure times due to the generation of thicker films and resultant lower cathodic efficiency.

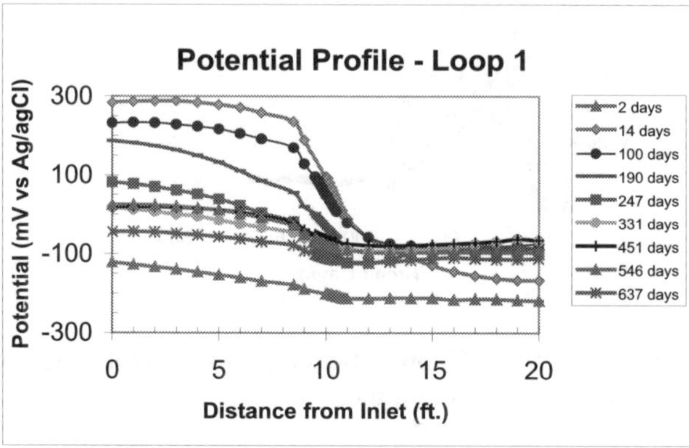

Figure 4: Titanium/70:30 Copper-nickel piping couple-potential profile of Loop No. 1 as a function of time.

4 Galvanic corrosion: modeling and engineering

Contact between dissimilar alloys will generate galvanic currents based on factors already delineated. The extent of risk to the overall systems and specifically to more electronegative alloy materials can be clarified if the magnitude of the corrosion rate at various locations can be modeled and calculated.

The potential distribution within a cylindrical system at a give time involves solution of the Laplace equation for the electrostatic potential in the seawater at any position given by cylindrical coordinates x (length along cylinder), and r (cylinder radius) [21]. Use of the Laplace equation assumes no concentration gradients in the seawater [22] and that all mass transfer results from ionic migration. Solution of the Laplace partial differential equation is non-trivial even for simple geometries and an analytical solution is not generally possible [21]. Fourier series solutions may be used for simple systems while finite element, finite difference, or boundary element analysis [21]. An alternative approach for simple geometry, cylindrical systems assumes that potential changes only in the longitudinal direction and that there is no radial component contributing to potential changes and that there is linear polarization kinetics [21]. An example of a symmetrical geometry for a dissimilar piping couple that is under cathodic protection in flowing seawater is shown in Figure 4. There the potential distribution is very nearly symmetrical with the asymmetry due the differences in open circuit potential of the Alloy A and the Alloy B pipes, respectively.

However, there are a number of examples where system/component geometry involving dissimilar alloys is not symmetrical or simple and the system may involve more than two dissimilar alloys. In addition, the contribution of flow is generally not considered. In systems containing elbows, the current and potential distribution will not be only in the longitudinal direction, but will have a radial contribution to current flow and corresponding potential changes.

Since oxygen reduction reaction is the predominate cathodic reaction, the rate of reaction may be limited by oxygen diffusion as described by:

$$i_c = 4FD_0C_{oxygen}/\delta$$

where i_c is the cathodic current density, F is Faraday's constant, D_0 is the diffusion coefficient in the electrolyte (seawater in this case), C_{oxygen} is the oxygen concentration in the bulk electrolyte, and δ is the thickness of the diffusion layer. The total ohmic potential drop in the electrolyte between any two points on the surface of the anode and the cathode can be calculated. The ohmic potential drop, ΔV_d, due to the distance between the anode and cathode is:

$$\Delta V_d = I_aR_d = I_aR_d$$

where $R_d = \rho d/\delta l$ with ρ the resistivity of the electrolyte, δ is the electrolyte thickness, 'd' is the distance between the anode and cathode, and 'l' is the width of the electrodes [10]. In the example where the more electropositive cathode is separated from the electronegative anode alloy by an elbow, there will be a variable distance between the anode and cathode leading to an asymmetrical current distribution. The distance between the anode and cathode along the elbow intrados, d_i, will be $d_i = ar_i$, where 'a' is a constant and r_i is the inner elbow radius. The cathode-to-anode distance, d_o, along the elbow extrados will be $d_o = ar_o \approx a(r_i + z)$ where r_o is the outside radius and z is the inside diameter of the elbow. As the diameter of the elbow increases, the disparity of anode-to-cathode distance along the inner radius verses anode-to-cathode distance along the outer elbow radius increases.

Under a thin layer electrolyte, the potential variation is large between the anode and the cathode. However, both the anode and cathode are only slightly polarized, except for the boundary between the cathode and the anode. With decreased fluid velocity, the electrolyte boundary thickness increases, and the galvanic current increases due to lower ohmic resistance. The current distribution become more non-uniform, and in the case where the anode and cathode are separated by an elbow, the galvanic current becomes concentrated along in the inner elbow radius plane since the current will follow a path of least ohmic resistance. Conversely, as the fluid velocity increases, the electrolyte thickness along the piping surfaces becomes thinner, increasing ohmic resistance resulting in a more uniform current distribution around the entire anode piping alloy and longer anode life.

5 Conclusions

The degradation of materials in marine environments depends on the expected materials demands and service requirements for the component or systems. Galvanic corrosion is dependent on a number of factors. Understanding these factors can optimize the ability of modelling to predict galvanic interactions in various environments and under various operating conditions.

References

[1] Oldfield, J.W., "Electrochemical Theory of Galvanic Corrosion," *Galvanic Corrosion*, H. P. Hack, ed., ASTM, West Conshocken, PA, 5-22, 1988.

[2] Francis, R., *British Corrosion Journal*, **29**, p. 53, 1994.

[3] Whitman, W.G., Russell, R.P., *Industrial Engineering Chemistry*, **16**, p. 276 (1924).

[4] Pryor, M.J., "Bimetallic Corrosion", Corrosion: Metal/Environment Reactions, vol. 1, 2nd Ed., L.L. Sheir, ed., Newnes-Butterworths, London, 1:192-214 (1976).

[5] Francis, R., *Galvanic Corrosion: A Practical Guide for Engineers*, NACE International, Houston, TX, pp.5-23 (2001).

[6] Hoxeng, R.B., Prutton, C.F., *Corrosion*, **5**, p. 330 (1949).

[7] von Fraunhofer, J.A., Lubinki, A.T., *Corrosion Science*, **14**, p.225 (1974).

[8] Hack, H.P., "Galvanic Corrosion of Piping and Fitting Alloys in Sulfide-Modified Seawater", Galvanic Corrosion, STP 978, H. P. Hack, ed., ASTM, West Conshohocken, PA, p. 339, (1988).

[9] Hack, H.P., "Atlas of Polarization Diagrams for Naval Materials in Seawater", CARDIVNSWC-TR-61-94/44 (April 12995).

[10] Zhang, X.G., "Galvanic Corrosion" in Uhlig's Corrosion Handbook, R. W. Revie, Ed., John Wiley and Sons, New York, pp. 154-160 (2000).

[11] Danek, G.J., *Naval Engineers J.*, **78**, p.763 (1966).

[12] Efird, K.D., *Corrosion*, **33**(1) 3 (1977).

[13] Gehring, G.A., Kuester, C.K., Maurer, J.R., CORROSION/80, 80032, NACE International, Houston, TX (1980).

[14] Gehring, G.A., Maurer, J.R., CORROSION/81, 81202, NACE International, Houston, TX (1981).

[15] Astley, D.J., *Corrosion Science*, **23,** p. 801 (1983).

[16] Scully, J.R., Hack, H.P., Galvanic Corrosion, ASTM STP 978, H.P. Hack, ed., ASTM, West Conshohocken, PA, pp.136-157 (1988).

[17] Shifler, D.A., "Possible Control Measures for Galvanic Corrosion in Piping Systems", Session 7, Corrosion Engineering I, Part 2, 1999 TriService Conference on Corrosion, November 15-19, 1999, Myrtle Beach, SC, DoD Tri-Service Corrosion Committee.

[18] Dexter, S.C., and Gao, G.Y., *Corrosion* **44**, p. 717 (1988).

[19] Eashwar, M., Maruthamuthu, S., Sathiyanarayanan, S., Balakrishnan, K., *Corrosion Science*, **25**, p. 1169 (1995).

[20] V. Scotto, M.E. Lai, *Corrosion Science*, **25**, p.1007 (1998).

[21] Astley, D.J., Galvanic Corrosion, ASTM STP 978, H.P. Hack, ed., ASTM, West Conshohocken, PA, pp.53-78 (1988).

[22] Newman, J. *Electrochemical Systems*, Prentice-Hall Inc., Englewoods Cliff, NJ, p. 232 (1973).

Polarization characteristics of geometrically confined spaces

M. Krupa[1], E. J. Lemeiux[1], A. Lueng[2], E. A. Hogan[2], V. G. DeGiorgi[2] & A. Seelinger[3]
[1]*Naval Research Laboratory, Key West, Florida, USA*
[2]*Naval Research Laboratory, Washington, DC, USA*
[3]*Naval Sea Systems Command, Washington, DC, USA*

Abstract

With the advent and application of Impressed Current Cathodic Protection (ICCP) systems to confined spaces, the ability of an externally controlled ICCP system to effect sufficient cathodic protection to internal components is questionable. This paper seeks to evaluate the ability of ICCP systems to provide adequate corrosion control for confined and sensitive surfaces. Both a boundary element and a physical model were utilized to understand the effects of current distribution within a confined space, through a 1.27 mm gap. Initially, tests were performed to quantify polarization levels in a simplified parallel plate configuration. Then more complicated geometries and materials were modeled to effectively evaluate real world scenarios. Previous tests have determined that adequate cathodic polarization is achieved for simple and complex geometries at the increased driving voltages experienced with sacrificial anode systems. The results of this work provide evidence that adequate cathodic protection may be provided at controlling potentials more electropositive to conventional galvanic cathodic protection system.
Keywords: confined, gap, cathodic protection, corrosion, impressed current.

1 Introduction

Physical Scale Modeling (PSM) has been used extensively by the Naval Research Laboratory (NRL) to determine appropriate positioning and size of ICCP system components on Navy vessels [1]. Anode and controlling reference cell positions are determined by analysis of potential profiles collected from the hull, via an array of reference cells placed in numerous hull positions and geometrically confined areas which have historically proven to be problematic. In addition, an analysis of current outputs from individual anodes is taken into account to provide system stability and meet controller design tolerances. Normal cathodic protection set potentials on a vessel's ICCP system controller will be -0.80 to -0.85 V vs. Silver-Silver Chloride (Ag/AgCl) reference cell. Through PSM testing conducted on numerous ship designs, variations in the cathodic polarization level deviating from the set potential due to hull geometry, damage conditions or electrolyte conditions have been identified.

Possible areas of concern are those that are at risk of underprotection due to the configuration of the hull and its appendages; namely, the propellers, rudders, struts, bilge keels, and components within ballast tanks. Physical scale models have been able to accurately identify problematic areas and additionally allow for solutions to many of these problems. This is primarily due to the fact that both thermodynamic-based and kinetic electrochemical effects on the electrode are inherent to the technique and accounted for in the modeling process. On the other hand, numerical methods require continuously updated iterations as a function of electrode surface conditions in order to obtain an accurate result [2].

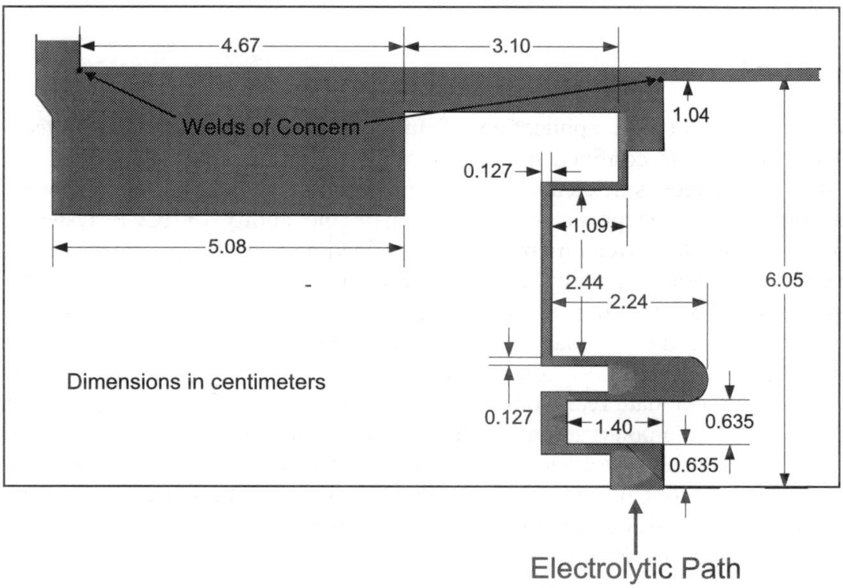

Figure 1: Cross-sectional depiction of the mating surface gap between concentric cylinders. Shaded region represents wetted portions.

In the current work, the level of protection of a confined and complex nickel aluminum bronze (NAB, UNS #95800) geometry that is enclosed in a ballast tank having an ICCP system was studied; a cross-sectional depiction of this geometry is shown in Figure 1. Adequate cathodic protection of these spaces is critical since they contain NAB welds which cannot be post-weld heat-treated; these welds are susceptible to stress corrosion cracking.

The gap considered in this effort is essentially a non-linear mating joint between to concentric cylinders approximately 1.5 m in diameter, fabricated from NAB with a weld material at the deepest portion of this gap. Previous ballast tank systems on earlier vessels, having this same confined geometry, had been protected with a galvanic zinc system, and under these conditions, this space received adequate protection as evidenced by the presence of calcareous deposits. Given the latter, no effort had previously been made to empirically characterize or analytically measure the electrochemical potential profile interior to this "gap" or confined space. Since new construction vessels are outfitted with ICCP in these spaces, it was undetermined whether sufficient cathodic protection could be anticipated. The hypothesis was that the zinc sacrificial system, generally results in a driving voltage of approximately 0.8V whereas the ICCP system will result in a driving voltage of 0.6V and therefore may not have enough throwing power to deliver sufficient current to the deepest portions of the NAB gap. This hypothesis was tested with the criterion that sufficient cathodic polarization was delivered when the deepest portion of the gap, such that a potential of –0.65 V vs. Ag/AgCl was achieved.

2 Phase I experiments: parallel flat plate physical model

2.1 Materials and methods

Although the electrolyte, and therefore the current from the ICCP system anodes, travels an intricate path through this particular confined space, it was decided to linearize this complex path for ease of testing and to simulate a best-case scenario. If the current could not penetrate into the gap through a straight, non-shielded path, then it was unlikely that actual conditions would permit adequate protection. Based on early evaluation and analysis of the drawings and pictures of the confined space, the first approach was to simply reproduce a simulated gap. The gap was to have the same configuration and dimensions as the actual structure, thus producing the same relative surface areas, as well as, electrolyte ohmic pathway. To reduce the experiment to a two dimensional problem, the gap was linearized by constructing a dual flat plate assembly having a depth equivalent to the path length of the original gap and a length equal to the circumference of the 1.2 m diameter concentric cylinders. The resultant configuration is shown in Figure 2. The minimum gap width noted on the original mating surface was 1.27 mm, and this was used as the separation between the flat plates.

The complete test assembly for Phase I is shown in Figure 3. Overall, Phase I testing was comprised of two 0.38 by 0.19 m flat plates of Copper Nickel 70/30 (CuNi 70/30, UNS # C71500), which has a similar cathodic polarization response to NAB. The plates were arranged in parallel and separated by a 1.27 mm gap, as detailed above. All external surfaces were coated with a MIL-P-24441 polyamide epoxy and three sides of the assembly were sealed. As a result, the wetted components were all interior to the gap and only one longitudinal face (shown in inset of Figure 2) allowed for current flow. Ten Ag/AgCl reference electrodes were mounted interior to the gap at ten locations along the gap length and at three gap depths so that potentials could be monitored. With regard to depth-wise reference electrodes, these were positioned at the mouth of the gap (Gap), at the middle of the gap (Gap centerline) and at the deepest point in the gap (Gap rear). In order to cathodically polarize the internal gap, a potentiostat controlled by an external reference electrode with a set potential of -0.8V vs. Ag/AgCl was positioned 0.46 m from the gap mouth as shown in Figure 3. Furthermore, a steel cathode material with a total surface area of 0.09 m^2, external to the gap, was included in the circuit to simulate actual ICCP conditions shipboard.

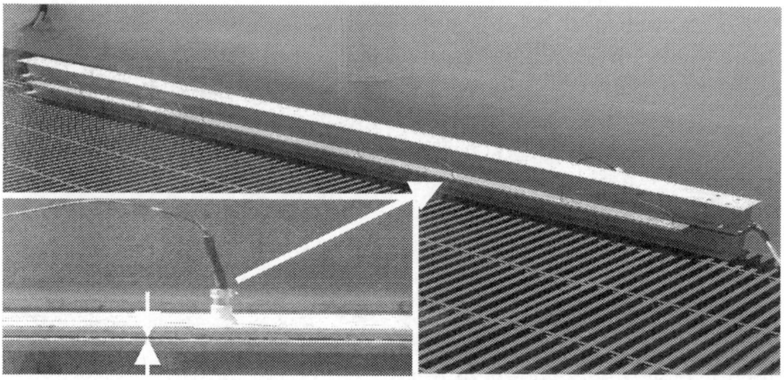

Figure 2: Pictorial of simulated gap flat plat assembly. Inset shows the mounted internal Ag/AgCl reference electrode and front face of internal 1.27mm gap.

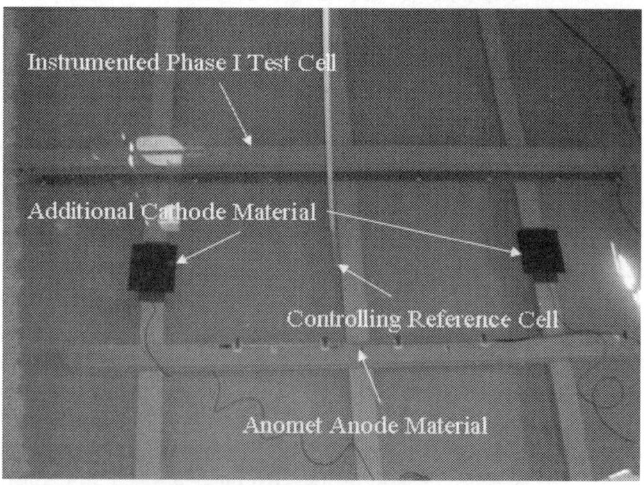

Figure 3: Pictorial of the test apparatus for Phase I testing.

2.2 Results

Internal gap potentials as a function of time are shown in Figure 4a. Anode potential was stable at approximately -1.5V vs. Ag/AgCl which is approximately half of the maximum design potential of this anode material. With regard to all potentials interior to the gap, potentials reached equilibrium after approximately 16 minutes of polarization. Note that the gap mouth (Gap) potential stabilized much sooner (6 minutes), than those deeper in the gap. This is clearly related to the increased resistance to current flow introduced by the narrow gap and subsequent formation of calcareous deposits. The potential subsequent to 16 minutes of testing for each location, depth wise, are plotted in Figure 4b. While these results indicate a depth wise differential in cathodic potential of 0.05 V, using the criterion established earlier, it was apparent that for this simplified linearized simulated gap, adequate polarization potentials were achieved at 0.19 m of penetration into the gap. This result indicated that testing of more complex geometries was warranted.

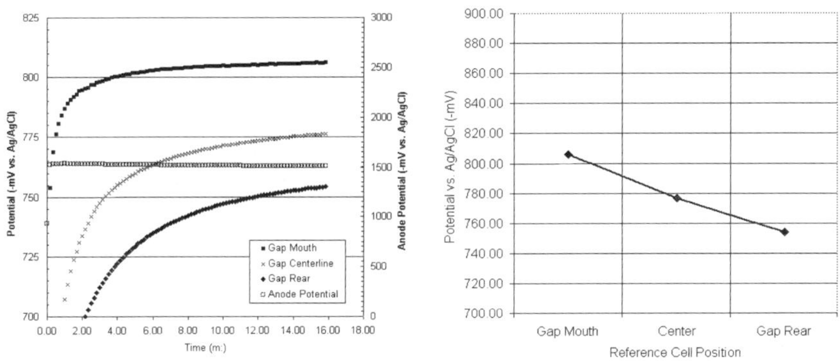

(a) (b)

Figure 4: Phase I testing results – (a) Gap potentials as a function of time and (b) Gap potential as a function depth.

3 Phase II experiments: physical model of complex gap geometry

3.1 Materials and methods

While Phase I tests did meet the -0.65 V vs. Ag/AgCl criterion, they also indicated a potential attenuation into the gap. Therefore, further potential attenuation was anticipated with the introduction of more complex gap geometry as considered in the Phase II testing. Given that the mating surface gap being considered was cylindrical and therefore symmetrical, a full scale, radial cross sectional, physical model was fabricated, as shown in Figure 5a. This test cell was then epoxy mounted with two 0.64 cm thick pieces of clear PVC as seen in Figure 5b. The model was instrumented with five Ag/AgCl reference cells in order to obtain the potential drop that would be seen upon penetration into this confined space. The complete configuration included, as before, a potentiostat and impressed current anode with an external controlling reference electrode.

The other major departure from the Phase I approach was that galvanostatic control was utilized instead of potentiostatic. From previous physical modeling of this ballast tank space [3], it was determined that the current density on these components was routinely between 75-140 mA/m^2 for all test situations. This was further validated since the design criterion used for that study required current densities of 108-161 mA/m^2 for a static condition. Given this realistic information, it was now possible to scale the current output from the galvanostat to these empirically derived current densities.

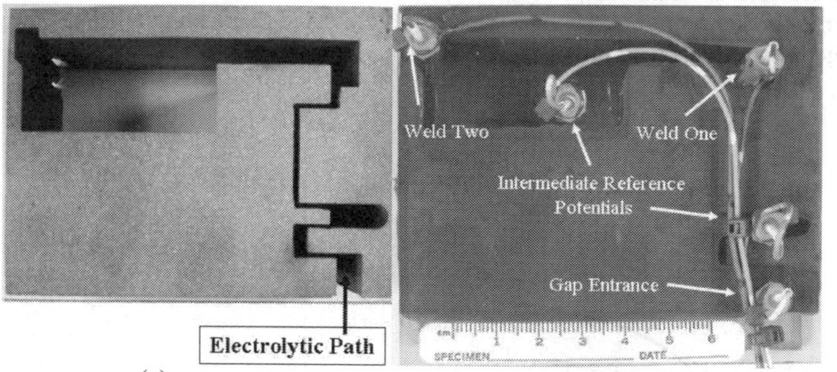

(a) (b)

Figure 5: Cross-sectional view of (a) internal mating surface of gap physical model and (b) reference electrode instrumented physical model.

Two different scenarios were to be considered: 1) the instrumented gap test cell in addition to an external cathode and 2) the instrumented gap test cell by itself. The surface areas for the external cathode and gap test cell were 0.123 m^2 and 0.023 m^2, respectively.

Following the previous ballast tank physical scale model results [3], a current density of 108 mA/m^2 was used to calculate the current output for two scenarios and these currents were 16 mA and 2.5 mA, respectively. However, during testing, it was evident that with the external cathode, 16 mA was insufficient to polarize the external cathodes to –0.80 V vs. Ag/AgCl in a reasonable amount of time, rapid polarization of components to set potential is ideal and this was not the case. When the test was initiated, the system came to steady state with the external cathode at -0.745V vs. Ag/AgCl. Therefore, the current was increased until the potential of the external cathode reference cell reached -0.8V vs. Ag/AgCl, which was achieved with a current output of 19 mA. Notably, this current output corresponded to a current density of 130 mA/m^2, which is within the aforementioned design criteria.

3.2 Results

The cathode potentials measured as a function of position within the gap are given in Figure 6. As with the simplified linear gap in Phase I, potential was attenuated with increasing distance into the gap. Note, however, the drop in potential was considerably higher with a total differential of approximately 0.125V which was a 2.5 fold increase over the previous test. This would indicate that the cathodic polarization of the internal mating surface was more a function of gap geometry than of the electrolyte resistance. Despite this attenuation, sufficient cathodic protection was achieved based on the original criterion of a minimum potential of -0.65V vs. Ag/AgCl. Since the weld location deepest in the gap had a potential electronegative to the design criterion, there was little risk of stress corrosion cracking at this location.

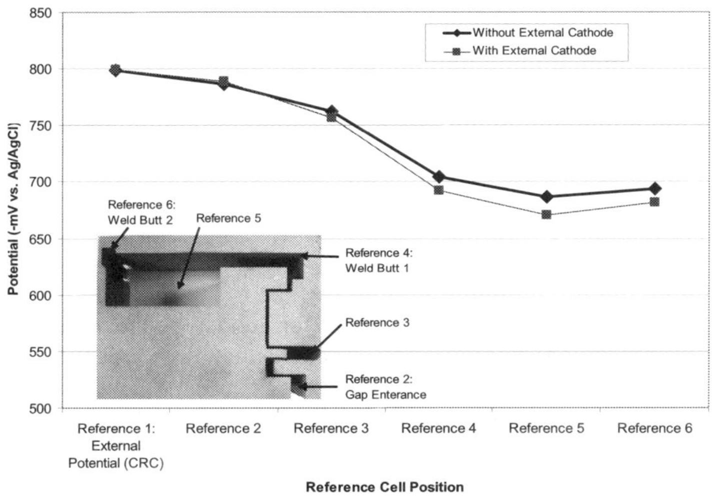

Figure 6: Cathodic potentials as a function of reference cell position of the gap model.

4 Boundary element modeling

It was decided that a mathematical or computational analysis of the enclosed space might provide additional insight into the phonological aspects of corrosion protection as observed in both the real structure and the physical scale model. Both finite element and boundary element computational techniques have been successfully applied to electrochemical processes. Because of the success of BE application to shipboard ICCP system analysis it was determined that initial analysis would be performed using BE techniques. The confined space was modeled using the commercial boundary element code BEASY [1]. The geometry modeled was identical to that used in the PSM testing.

The confined space surfaces were assigned NAB material properties derived from a NAVSEA defined polarization curve. The potential-current density polarization response was shifted to match observed experimental results. The shape of the experimentally determined curve is maintained. However, the curve is shifted so that current density corresponds to that measured during PSM at the reference set point. The original laboratory method NAB polarization response and shifted polarization response are shown in Figure 7.

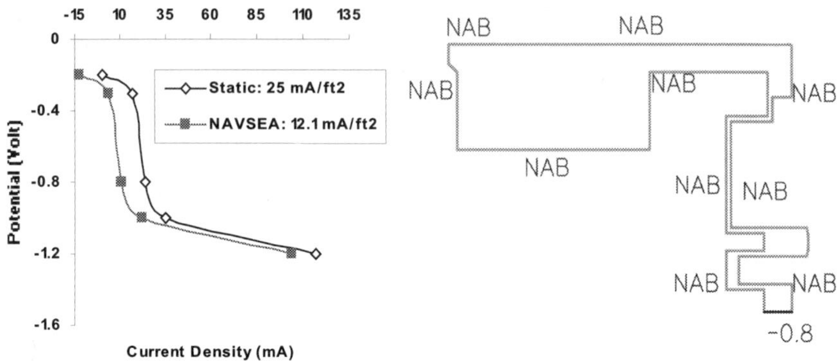

(a) (b)

Figure 7: (a) Polarization curve used for model boundary conditions (b) boundary model used to solve potential distribution.

The boundary conditions applied to the boundary element model are NAB polarization response applied to all but the section representing the gap entrance. The gap entrance is assigned a voltage value of –0.8 V vs. Ag/AgCl. The region internal to the confined space was assigned an electrolyte conductivity equivalent to that used in PSM.

The resulting potential map of the internal volume of the confined space is shown in Figure 8. As can be seen, the potential variation between the gap entrance and the enclosed region is much larger than that observed in either experimental procedure. Almost no current reaches the larger volume. The extreme difference in calculated and experimental results was unexpected and lead to a review of the basic assumptions of the BE technique.

BE techniques as applied to this problem assume steady state electrochemical corrosion governed by LaPlace's equation:

$$k\nabla^2\Phi = 0 \tag{1}$$

Where Φ is the electric potential and k is the conductivity of the electrolyte within a domain. There are three necessary conditions for the valid application of LaPlace's equation. There can be no gain or loss in electrical current in the system. There can be no sources or sinks within the defined domain. These conditions were met. The electrolyte must be of a uniform composition. It is felt that this is not the case in the confined space geometry. There are stagnant regions, flow is constrained and regions of the electrolyte volume are essentially isolated from other regions with little mixing between the volumes. Therefore it appears that the confined space geometry does not lend itself to LaPlace based BE techniques. While the LaPlace solution based BE technique may be a reasonable approximation for a ship in the open sea and in deep water it is not appropriate for the confined space geometry in this work. Based on this no further comparisons of results were made.

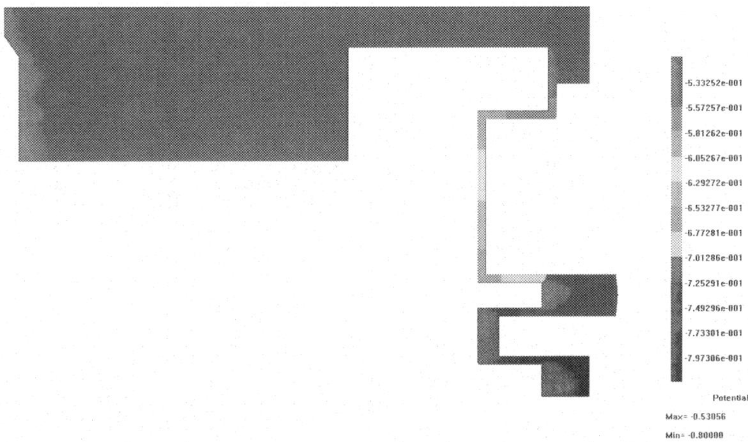

Figure 8: BEM result of potential distribution interior to the gap.

5 Summary

The results from the Phase I testing was utilized to determine if it was necessary to model the complexity of the gap in question. The parallel flat plate gap testing revealed that electrolytic resistance alone was insufficient to appreciably attenuate cathodic protection potentials below -0.65V vs. Ag/AgCl in a 0.19 m deep gap when an exterior cathode is controlled to -0.8V vs. Ag/AgCl.

As a result, subsequent testing was conducted (Phase II) which included both the effects of electrolytic resistance and surface geometry/shielding effects. The latter was accomplished through the use of a radial slice, physical model of the actual in-service mating surface assembly. Additionally, the experiment was operated in galvanostatic mode with current outputs delineated from previous half-scale physical scale modeling of this ballast tank space. The results of these instrumented gap test, demonstrated a potential drop of between 0.1 and 0.15 V can be expected upon complete penetration into the gap. Clearly, the geometry of the gap affected the potential attenuation than did the solution resistance. Despite the resultant attenuation, however, the NAB weld metal, internal to the gap assembly, remained electronegative to the protection potential of -0.65 V vs.

Ag/AgCl. Thus the physical model indicates that sufficient cathodic protection is available to prevent stress corrosion cracking of NAB weld metal and to protect the internal surfaces of the mating assembly.

While the application of BE techniques to the confined space was not successful it provided essential insight into the limitations of the computational method as it currently exists in commercially available codes. The confined space does not appear to meet one of the fundamental requirements for use of LaPlace's equation; specifically the electrolyte is not of uniform composition. There is a need for computational methods that are valid for volumes with extremely non-uniform electrolytes. This need will increase in the future as more ICCP systems are installed in confined spaces. It is hoped that future research will provide the computational tools needed by the design community.

References

[1] Lucas, K.E., Thomas, E.D. & Hogan, E.A., "Theory and Application of Physical Scale Modeling (PSM) for the Design of Impressed Current Cathodic Protection Systems," InterNaval Corrosion Conference, Monterey, CA, April 1992.

[2] Munn, R.S. & Devereux, O.F., "Numerical Modeling and Solution of Galvanic Corrosion Systems: Part II. Finite-Element Formulation and Descriptive Examples," *Corrosion*, August 1991.

[3] K. E. Lucas, M. A. Evans, E. A. Hogan and E. D. Thomas "Criteria for the Design of Impressed Current Cathodic Protection (ICCP) for Seawater Ballast Tanks" Tri-Services Conference Proceedings, Nov 1997.

[4] BEASY User's Manual, Computational Mechanics Ltd., Billerica MA, 2003.

Calculation of temperature fields with DC and pulsed ECM

N. Smets, G. Nelissen, J. Deconinck & S. Van Damme
Department ETEC, Vrije Universiteit Brussel, Belgium

Abstract

During the process of Electrochemical Machining (ECM) metals are electrochemically dissolved at very high current densities up to 1 A/mm^2. Simulation of the temperature distribution during the process provides interesting insights and guidelines for practical use. To be able to calculate the temperature distribution a coupled model with a time accurate solver (second order accuracy) is implemented. In practice, a pulsed current is applied for reasons of accuracy and surface quality. Hence the calculation of the temperature field during pulsed ECM as a function of time is performed, but this can be a very expensive procedure. The pulses have to be described on a timescale that can be orders of magnitude smaller than the timescale on which the thermal effects evolve. If one wants to calculate the temperature evolution one must calculate the effect of all of these pulses. A new approach is introduced. By replacing the pulsed current with a constant current, which has the same thermal effect, it is shown that one can calculate the temperature field in a much more efficient way. The calculation is done with large timesteps, which is a great improvement. Results of expensive simulations with pulsed current and cheap simulations with thermally equivalent constant current are performed and shown to be very comparable.
Keywords: pulsed ECM, PECM, temperature, heat, electrochemical simulation.

1 Introduction

Electrochemical machining (ECM) is an important manufacturing technology in machining difficult-to-cut materials and shaping complicated contours and profiles. ECM is the controlled electrochemical dissolution of a metal. In an ECM cell one can distinguish two electrode types: the workpiece (anode), and the tool (cathode). Between the electrodes there is a small gap through which an electrolyte is being pumped. During the ECM process a low voltage is applied between the electrodes, producing a high current density (0.1 – 1 A/m^2).

Pulsed Electrochemical Machining (PECM) involves the application of current or voltage pulses instead of DC current or voltage in the ECM process. One wishes to apply pulsed current for reasons of accuracy and surface quality [2].

Simulation of the temperature distribution during the process provides interesting insights and guidelines for practical use. Experimental and theoretical results indicate that the ECM process should be conducted at the highest possible current density to maximize the accuracy and the surface finish. The issue of heating of the electrolyte is of primary importance for the determination of the limit conditions in ECM [1].

2 The model

The full electrochemical model focussing on the isothermal electrochemical processes is subject to a separate paper [3]. In this paper only thermal aspects are considered.

2.1 Conservation of charge

The current density is governed by the Laplace equation

$$\overline{\nabla}.\overline{J} = \overline{\nabla}.\left(-\sigma(T)\overline{\nabla}U\right) = 0 \tag{1}$$

with non-linear boundary conditions due to the electrochemical reactions. Although in the full model the conductivity σ(T) is also a function of the concentrations of the iron and hydroxide ions, for short electrodes these effects can be neglected. In what follows σ(T) = 16.44(1+0.02(T-293K)) S/m.
Further a linear overpotential is being used, as in eqn(2).

$$J = a\eta + b \tag{2}$$

The values for a and b which are used in this work are given in table 1.

Table 1: Parameters of the linear overpotentials used in this work.

	a (A/V/m^2)	b (A/m^2)
Anode	2.91E5	-3.31E5
Cathode	1.45E7	1.02E7

2.2 Electrolyte flow, velocity field

In this work the incompressible Reynolds averaged Navier-Stokes equations are solved in order to find the local flow field. The turbulent viscosity which enters these equations, is calculated with the low-Re k-ω turbulence model. A detailed description of the equations, the boundary conditions and the numerical solution techniques can be found in [5]. It is considered that the metal removal rate is small compared to the flow rate. Hence no electrode shape change has to be considered, so that the flow may be regarded as stationary over several ECM pulses.

2.3 Conservation of heat

Heat conservation is considered in the electrolyte and in both electrodes. In the electrode, diffusion and ohmic heat production are taken into account. In the electrolytic solution convection, diffusion and ohmic heat production are taken into account. The following coupled equations are solved:

$$\frac{\partial T}{\partial t} = \overline{\nabla}.\left(\alpha_{electrode}\overline{\nabla}T\right) + \frac{\alpha_{electrode}}{k_{electrode}}\rho_{electrode}J^2(t) \tag{3}$$

in both electrodes, and

$$\frac{\partial T}{\partial t} + \overline{\mathrm{v}}_{\mathrm{solution}}.\overline{\nabla}T = \overline{\nabla}.\left(\alpha_{\mathrm{solution}}\overline{\nabla}T\right) + \frac{\alpha_{\mathrm{solution}}}{k_{\mathrm{solution}}}\rho_{\mathrm{solution}}J^2(t) \tag{4}$$

in the solution, α being the thermal diffusivity (m^2/s), k the thermal conductivity (W/mK) and ρ the electrical resistivity (Ωm). A significant amount of heat is being produced at the electrode interfaces where the electrochemical reactions take place. Therefore at both electrode surfaces the local heat sources $\eta(t).J(t)$ (W/m^2) are considered.

2.4 Numerical method

All partial differential equations presented above are solved in two dimensions using the residual distribution method [6]. The scalar N-scheme is applied to the convective terms in the convective diffusion equations. Standard finite element method (FEM) discretization is applied to the diffusion terms. According to [4] FEM is the most comprehensive method to simulate ECM. All the numerical schemes provide at least second order accuracy [7]. An implicit second order time accurate time integration scheme is used [8].

3 Thermally Equivalent Constant current (TEC current)

The straightforward calculation of the temperature field as a function of time can be performed, but is an expensive procedure. The current evolves on a timescale (ms) that can be orders of magnitude smaller than the timescale on which the global thermal effects evolve (s). This means that, to calculate the evolution of the temperature, one needs to calculate a lot of timesteps, needed to take into account the full detail of the pulses.

A new approach is introduced to tackle this problem. By replacing the pulsed current with a constant current, which has the same thermal effect, it is shown that one can calculate the temperature field in a much more efficient way. The calculation can be done with large timesteps, which is a great improvement.

In the doublelayer at the anode one doesn't have a constant resistance, since that resistance is a function of the current density. The average heat generated by one pulse during a period T is

$$Q_{average} = \int_0^T \eta(t)J(t)dt = \eta_{pulse}J_{pulse}\alpha T\,, \tag{5}$$

where α is the duty cycle of the pulse, which is the length of the on-time divided by the period. One would like that the TEC current density J_{TEC} produces the same amount of heat during that time such that:

$$Q_{average} = \int_0^T \eta(t)J(t)dt = \eta_{TEC}J_{TEC}T\,. \tag{6}$$

A linear polarisation (as in eqn (2)) will be used here, but the same reasoning can be generalised and applies e.g. also to the Butler-Volmer polarisation. After a small calculation one finds that

$$J_{TEC} = \frac{b}{2} + \frac{1}{2}\sqrt{b^2 + 4\alpha(J_{pulse}^2 - bJ_{pulse})}\,. \tag{7}$$

In case b is zero, the doublelayer has a constant electrical resistance, and eqn (7) reduces to

$$J_{TEC} = J_{pulse}\sqrt{\alpha}\,, \tag{8}$$

which is the expression for the RMS value of a pulse. It is easily verified that if $\alpha = 1$ eqn (7) reduces to

$$J_{TEC} = J_{pulse}\,. \tag{9}$$

In the doublelayer at the cathode one has the same situation, but the TEC current is different from that at the anode, because there is a different polarisation behaviour.

Finally the TEC current in the electrolyte and the electrodes is simply the RMS value of the pulsed current (e.g. constant resistivity).

The same current flows through both the doublelayer at the anode and cathode , the electrolyte and the electrodes, but both the doublelayers have different TEC currents than the other locations. As a consequence one can't make all the locations thermally equivalent, while applying only one current. The most significant heat source will be made thermally equivalent, which is the doublelayer at the anode.

So the thermally equivalent constant current (TEC current) is being defined as the constant current, which generates the same average heat in the doublelayer at the anode than the pulsed current, when injected in the electrochemical system. One should handle this thermally equivalent constant current with care: the same average heat production in time does not necessarily mean that the same average temperature field in the system is established, but the presented approach will show to be very applicable.

4 Results and discussion

A simulation is performed on a short channel, such that the steady state is reached after only a few pulses.

4.1 Geometry

The geometry on which calculations are performed is a simple channel. This channel is 0.208 mm high, and 7.4 mm in depth (perpendicular to the 2D surface). The anode is 10 mm long and 0.5 mm high. The cathode is 5.06 mm long and 0.5 mm high. Both electrodes are centered. A scaled drawing is shown in figure 1. The electrolyte flows with an average velocity of v_{flow} = 15 m/s, and enters fully established the reactor at a temperature of 293 K.

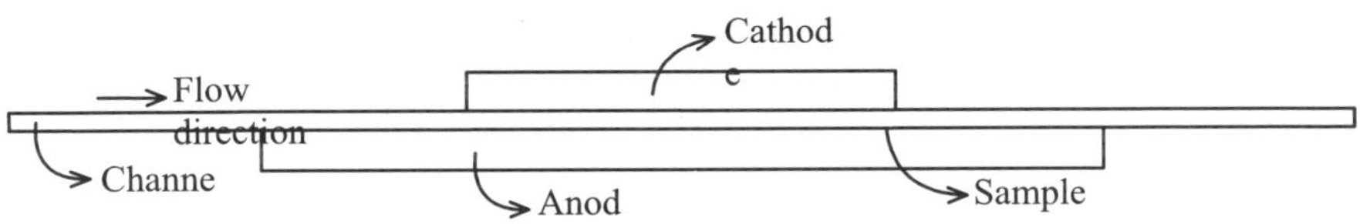

Figure 1: Geometry of the short channel.

The outer sides of the electrodes are considered thermal isolators.

In order not to have to consider and compare whole 2D and time dependent temperature fields, one temperature is taken from a sample point. This sample point gets chosen in the hottest spot in the channel, since that is the major point of interest. This hottest spot is the most downstream point where current flows, in the thermal layer that builds up at the anode (x = 0.0017754 m ; y = 0). An example of a calculated temperature field is shown in figure 2.

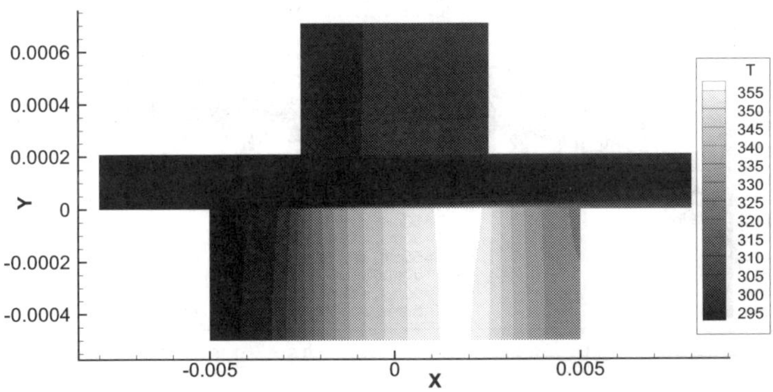

Figure 2: Example of a temperature field, T in K, X an Y in m, t=10s, α =100%, mind the different scales for X and Y.

4.2 Applying pulses with equal TEC current, but different duty cycles

Pulses with duty cycles of 10%, 25%, 50%, 75%, 90% and 100% are applied. All the current pulses have the same TEC current. The period T is 50ms.

When applying the currents listed in table 2, one can follow the evolution of the temperature in the sample point as a function of time as shown in figure 3.

The smaller the duty cycle, the shorter the pulses, the higher the current density during the pulse and the higher the temperature peak in the sample point.

Each curve consists of two components: a slow transient and a ripple because of the pulses. One can notice a comparable evolution of the slow transients in the cases with varying duty cycles, which means that the concept of the TEC current is applicable in this case. In the end a quasi-steady state is reached, and a true steady state for the simulation where DC current is applied. By applying the TEC current one calculates the average evolution of the temperature. The ripple however, causes large peaks in the temperature evolution which are important for determining the process limits.

Table 2: Applied currents.

α	I_{on} [A]	J_{on} [A/mm^2]
10%	80.57	2.15
25%	48.89	1.31
50%	33.00	0.881
75%	26.01	0.695
90%	23.31	0.623
100%	21.86	0.584

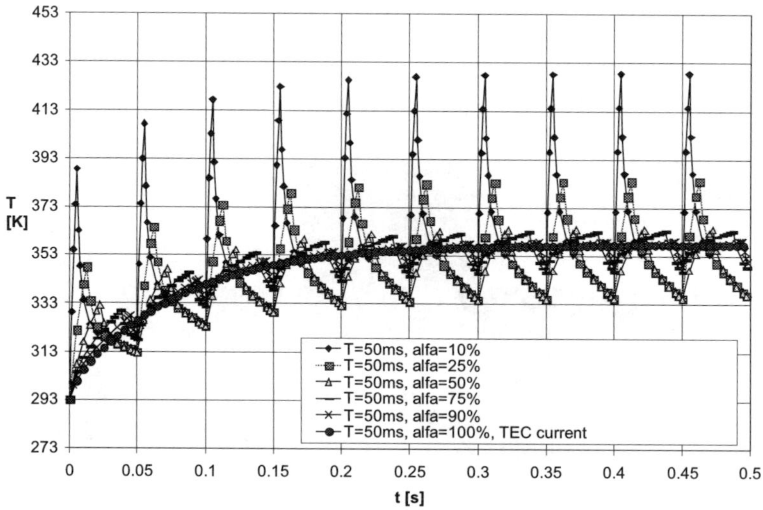

Figure 3: Evolution of the temperature in the sample point as a function of time, pulses with different duty cycles and constant TEC current applied.

4.3 Quasi-steady state shortcut

Using the TEC current the steady state can be reached after only a few timesteps. At this steady state, one can apply a few pulses to obtain the quasi-steady state temperature evolution.

In what follows three different currents are applied to the electrochemical system. First, TEC current will be applied during the slow transient, and afterwards pulses. The slow transient can be passed with only a few timesteps as in figure 4 (left), or more timesteps can be used to get a better result from the calculation during the slow transient as in figure 4 (right). To be able to compare the results also a full pulsed case is calculated as in figure 5.

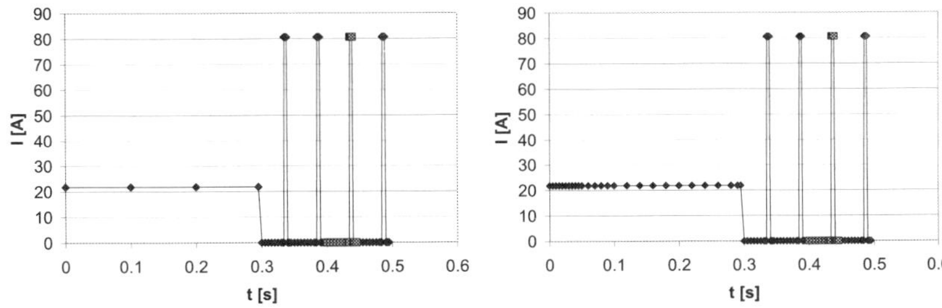

Figure 4: Applied current with TEC current during slow transient and pulses afterwards. Left graph with few timesteps, and right graph with many timesteps during the slow transient.

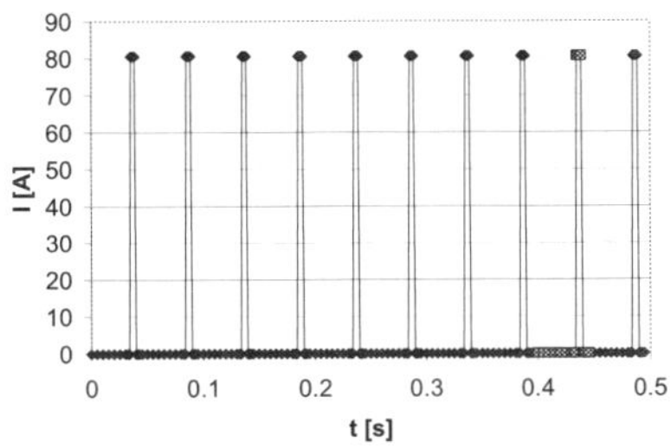

Figure 5: Applied current, only pulses.

4.3.1 Analysis of the temperature in the sample point

The resulting temperature in the sample point as a function of time is shown in figure 6.

In figure 6 one can see that using many timesteps while applying TEC current during the slow transient doesn't pay off, using only a few timesteps brings us in almost the same state. Passing the slow transient using only a few timesteps is a very cheap operation. Calculating a few pulses is also an inexpensive operation. Applying this method, the temperature field on the whole geometry during the whole quasi-steady state, can be obtained with a minimum of calculations.

In this case the slow transient doesn't take a long time because of the small volume of the workpiece, but the slow transient can be very long. For instance, when the workpiece has a large volume, the duration of the slow thermal transient can be orders of magnitude larger than the pulse period of the current.

4.3.2 Analysis of the 2D temperature field

The use of the sample point is a way of simplifying the visualisation, but still there is a whole 2D field that one has to consider. In order to have a quantitative measure for the quality of the 2D field, the procentual error $\Delta_{\%}$ between 2 fields is calculated as shown in eqn (10). T_0 is the initial temperature field where T = 293K in every point of the geometry. The area of calculation of $\Delta_{\%}$ is restricted to the points where the temperature rise is significantly large, e.g. where $|T_1-T_0|$ is greater than 10% of the maximum difference between T_1 and T_0.

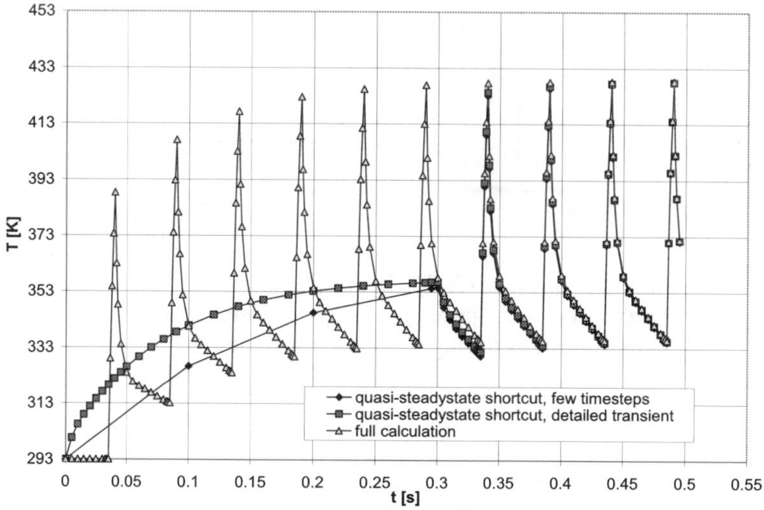

Figure 6: Evolution of the temperature in the sample point as a function of time, quasi-steady state shortcut.

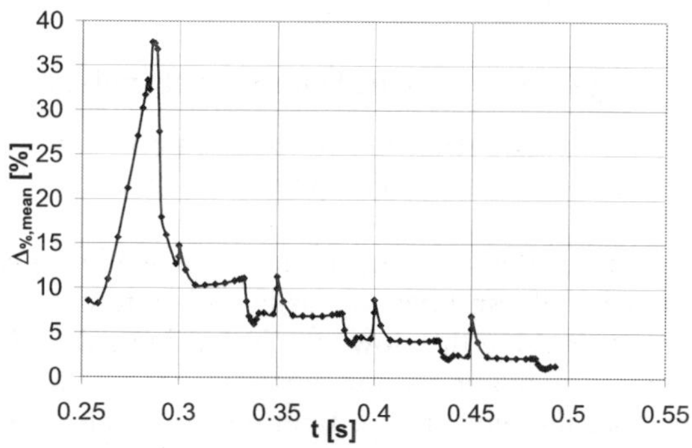

Figure 7: The mean procentual error $\Delta_{\%,mean}$ between the case with the quasi-steady state shortcut with few timesteps during the slow transient, and the case where pulses are applied all the time.

$$\Delta_{\%}(x,y) = \frac{|T_2(x,y) - T_1(x,y)|}{|T_1(x,y) - T_0(x,y)|} 100 \quad . \tag{10}$$
$$\forall x, y \;\middle|\; |T_1(x,y) - T_0(x,y)| > 10\% \max(T_1 - T_0)$$

Further the mean procentual error $\Delta_{\%,mean}$ is defined as follows

$$\Delta_{\%,mean} = \frac{\iint_S \Delta_{\%}(x,y)dS}{\iint_S dS}, \tag{11}$$

with S being the area of the geometry where $\Delta_{\%}(x,y)$ is defined.

The mean procentual error $\Delta_{\%,mean}$ between the case with the quasi-steady state shortcut (with few timesteps during the slow transient), and the case where pulses are applied all the time is shown in figure 7.

It is observed in figure 7 that from the moment pulsing is applied ($t = 0.335s$) $\Delta_{\%,mean}$ drops below 10%, which is already very good, considering that the procentual error on the thermal parameters may easily be 10%. Even if the value of the TEC current would be poorly chosen, the quality of the quasi-steady state shortcut can still be good, because one could apply a convergence criterion. The longer the pulses are applied, the better the match becomes.

5 Conclusions

Results of expensive simulations with pulsed current and cheap simulations with thermally equivalent constant (TEC) current are performed and show to be very comparable. The calculation of the temperature in the system can be simplified a lot by using the TEC current, which allows to use larger timesteps.

One can pass the slow thermal transient by applying TEC current with only a few timesteps. Calculating a few pulses afterward is an inexpensive operation. Applying this method, the temperature field of the whole geometry during the whole quasi-steady state is obtained, with a minimum of calculations.

References

[1] Kozak, J., Rajurkar, K.P., and Lubkowski, K., "The Study of Thermal Limitation of Electrochemical Machining Process", Transactions of the NAMRI/SME, Vol. XXV, pp. 159-164, 1997.

[2] Kozak, J., and Rajurkar, K.P., "Computer Simulation of Pulse Electrochemical Machining (PECM)", Journal of Materials Processing Technology, Vol. 28/1-2, pp. 149-157, 1991.

[3] Van Damme, S., " Numerical model for predicting the efficiency behaviour during pulsed electrochemical machining of steel in $NaNO_3$", Journal of applied electrochemistry, submitted to be published

[4] Jain, V.K., Rajurkar K.P., "An Integrated Approach for Tool Design in ECM", Precision Engineering, Vol. 13, No. 2, pp. 111-124, 1991.

[5] Nelissen, G., Van den Bossche, B., Deconinck, J., Van Theemsche, A. and Dan, C., " Laminar and turbulent mass transfer simulations in a parallel plate reactor", Journal of Applied Electrochemistry, Vol. 33, pp. 863, 2003.

[6] Nelissen, G., Van Theemsche, A., Dan, C., Van den Bossche, B. and Deconinck, J., " Multi-ion transport and reaction simulations in turbulent

parallel plate flow", Journal of Electroanalytical Chemistry, Vol. 563, pp. 213, 2004.

[7] Waterson, N., "Simulation of Turbulent Flow, Heat and Mass Transfer Using a Residual-Distribution Approach", Dissertation, Technische Universiteit Delft, 2003.

[8] Dan, C., Van den Bossche, B., Bortels, L., Nelissen, G. and Deconinck, J., "Numerical simulation of transient current responses in diluted electrochemical ionic systems", Journal of Electroanalytical Chemistry, Vol. 505, pp. 12, 2001.

Durable building technology benefiting electrochemical methods as a preventive

D. A. Koleva, J. Hu & A. L. A. Fraaij
*Faculty of Civil Engineering and Geosciences,
Delft University of Technology, The Netherlands*

Abstract

Two series of Portland cement mortar prisms were cast with variations in experimental set-up and chloride concentration. All samples were subject to resistance monitoring, ion concentrations measurements and pore structure analysis. The research reveals minor influences on mortar resistivity of changes in chemical composition associated with ion diffusion and migration for the investigated cases. The electrical resistivity was demonstrated to be mainly dependent on the microstructural changes induced by hydration and chemical binding mechanisms.
Keywords: electrical resistivity, chloride concentrations, pore structure, concrete corrosivity.

1 Introduction

This contribution presents part of the research outcomes of a comprehensive study on corrosion prevention and protection by electrochemical methods (conventional and pulsed cathodic protection). The efficiency of these electrochemical protection techniques (based on applied electrical current) depend on, among other things, electrical resistivity of the electrochemical system of concrete-steel-pore solution. Thus concrete resistivity and conductivity are related to the corrosion process on one hand and the proper execution of electrochemical protection techniques on the other. A target of this preliminary study is to investigate the influence of ion movement (resulting from concentration gradient) on the resistivity of cement-based porous systems in rest conditions. This study additionally pursues to explore the morphological changes of material microstructure, particularly of pore structure, induced by the ion movement, ion binding and resistivity.

2 Experimental

2.1 Materials and experimental design

Mortar prisms of 40x40x160 mm were cast according to standard experiment procedures. The mixing proportions are 1:3 for Portland cement CEM III 32.5 and sand (size range between 125 μm and 2 mm), with water-to-cement ratio (w/c) of 0.5. The specimens were cured in fog room (at 20°C) with relative humidity (RH) of 98% during the whole period of testing. Three groups of mortar specimens were made, varying in chemical composition and experimental set-up. The first group was free of additives (T series), the second was produced using 3% NaCl solution as mixing water (N series) and the third one comprised of half-to-half combination of the first two groups and denoted as TN series. Stainless steel pins were equidistantly embedded (into 30 mm depth) in the longitudinal direction of the specimens for resistance monitoring; the specimens were additionally subjected to determination of chloride and alkali concentrations and pore structure investigation.

2.2 Electrical resistivity measurements

Resistance monitoring was conducted on specific sections of the specimens to provide insight into ion concentrations gradient due to diffusion. Measurements of the total resistance gave information on the changes of electrical properties in the specimens with various aggressive ion concentrations and thereby under concentration gradients. The resistance measurement was performed by DC 4 pin and AC - 2 pin and 4 pin methods. For the DC measurement, a DC current of 2mA was applied to the outer pins 1 and 5 (scheme in Fig. 1) and the voltage drop in every section was recorded. For calculating total resistance the voltage drop between pins 2 and 4 was taken as a value. For the AC measurement (2 and 4 pin) a specially designed R-meter was used. By applying alternating current of 1mA at frequency of 1kHz, the device outputs a DC voltage, which represents the actual resistance. The different methods yielded similar outcomes (the coefficient of variation was within 3 to 5%) and showed the same trend of changes in resistance. The electrical resistivity was calculated on the basis of the recorded resistance values via the equation: $\rho = RA/l$, where ρ is the resistivity in Ohm·cm, R is the resistance in Ohm, A the cross section in cm^2 and l the length in cm. This equation was considered suitable for the present set up in view of the geometry of the specimens as well.

2.3 Chemical analysis

Chemical analysis was performed for the selected specimen sections at 1, 8, 14, 24, 30, 41 and 57 days of hydration. The analysis aimed at monitoring ion transport in rest specimens resulting from concentration gradients. Chloride concentrations were determined according to Volhard titration method, while cation concentrations were measured by means of plasma spectrometry using

Inductive Coupled Plasma spectrometer (ICP-EAS). Chemical analysis was performed in agreement with ASTM C114-77.

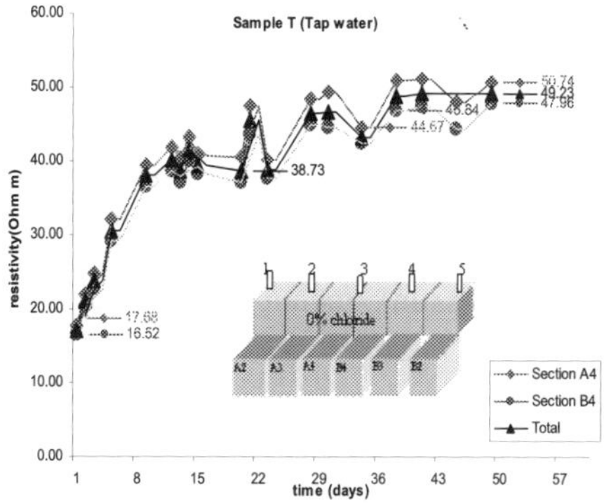

(a) Sample T (using tap water)

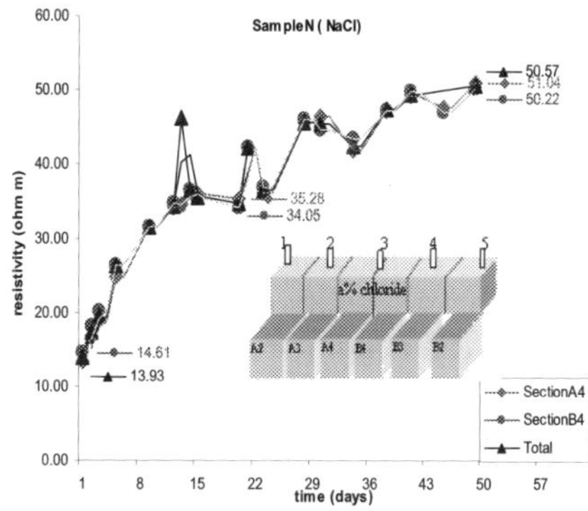

(b) Sample N (using 3%NaCl)

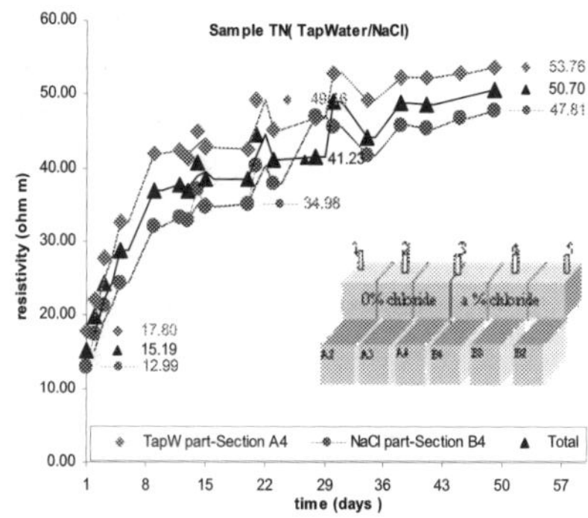

(c) Sample TN ('mixed sample')

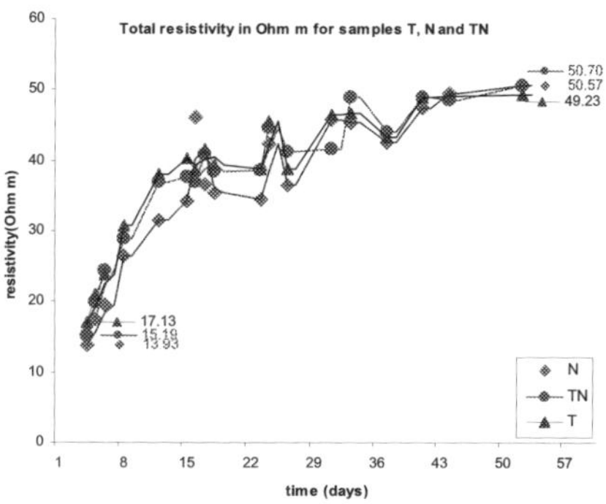

(d) Samples T,N and TN

Figure 1: Mortar resistance versus time.

3 Results and discussion

3.1 Electrical resistivity

The mortar resistivity increases with time, showing similar trend for all groups of samples. In the beginning of the test, the samples with NaCl showed slightly lower values compared to the samples made with tap water and the sample tap water/NaCl (the 'mixed sample' TN). At the end of the test - approximately 60 days, all samples had almost equal values, despite the significant difference in ion concentrations. When the hydration becomes a diffusion-controlled process, it strongly depends on the mobility of ions and water in the pore solution, i.e. on the speed of hydration products formation and of filling up the pore space with these hydration products. Additives as NaCl are known as accelerators of the hydration process, thus it could be expected that the hydration process exerts less influences on the electrical resistivity in the T samples than in the case of N samples. It was claimed in other investigations [1] that the presence of chloride ions, especially free chloride (chloride in the pore solution) will lead to lower electrical resistivity. This is deviating from what is observed in this study (see Fig. 1).

The N samples showed lower resistivity values up to 35 days of hydration and arrived at similar values with the T sample till the 60th day of hydration (Fig. 1d). Thus, at the end of the test the total resistivity of the different samples was almost equal, the highest being for the 'mixed' TN sample. It can be concluded that the electrical resistivity is not dependent on the ion concentration and increasing resistivity does not mean low aggressive ion concentration, especially for high RH conditions.

3.2 Chloride and alkali concentrations

Chloride (free and total chloride) and alkali concentrations were determined for each section of all the prisms. It was expected that due to concentration gradient in TN specimens, chloride ions will migrate from higher concentrations zone (part B) to lower concentrations zone (part A) of the specimens, thus influencing the electrical properties of the mortars. The same holds for sodium ions. However the global diffusion process was very slow under the present testing conditions.

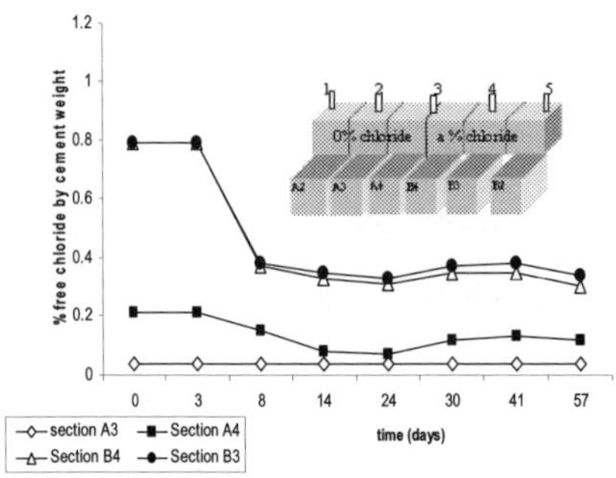

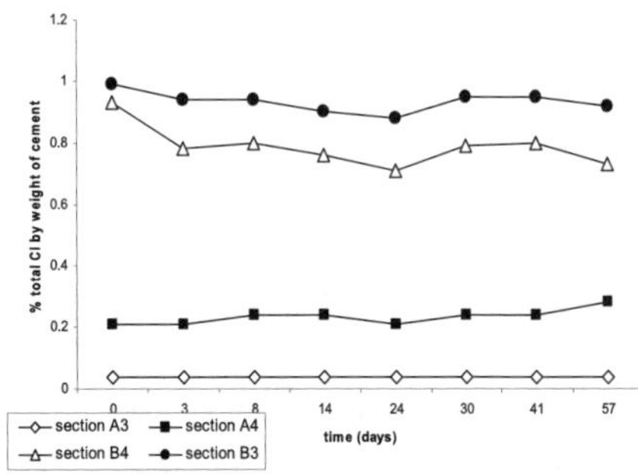

Figure 2: Chloride concentration (% of cement weight) versus time for free Cl^- (left) and total Cl^- (right) in TN.

Fig. 2 shows constant ion concentration in section A3 of the TN samples; hence, chloride ions did not migrate into that part for the period of experiments. In section A4 a slight diffusion-induced increase in Cl^- concentrations is recorded. The starting point chloride concentration was 0.2% by cement weight (free chloride equals total chloride since there was still no binding). The total chloride concentrations slightly shifted towards higher values and reached about 0.3 % at 60 days of hydration. The free chloride concentration is decreasing due to the chemical chloride binding during the ion diffusion process. The similar patterns of changes but in an opposite direction are observed for sections B3 and B4 of the samples TN, indicating a very slow diffusion process. The chloride concentrations are much higher in part B of the specimens than those in part A, accompanied by a decline from 0.93 % to 0.73 % in total chloride concentration and from 0.79% to 0.3 % in free chloride concentration (Fig. 2). With respect to mortar corrosivity, the chloride concentration in the part B of TN samples was sufficiently high to cause corrosion if reinforcements were present [2]. The risk is existent even in section A4 of the samples with significantly lower chloride concentrations.

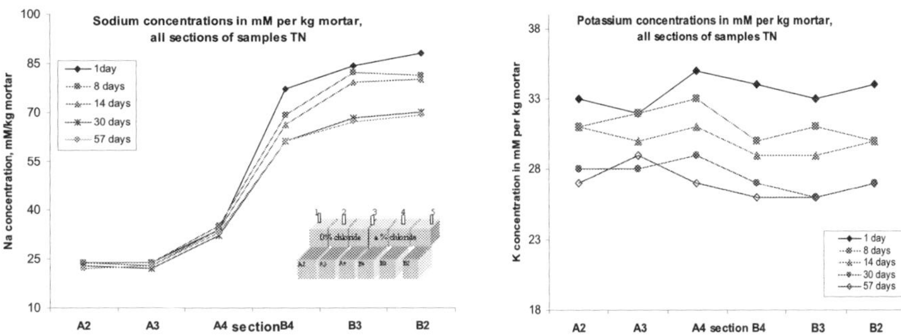

Figure 3: Concentration of Na^+ (left) and K^+ (right), expressed in mM per kg mortar.

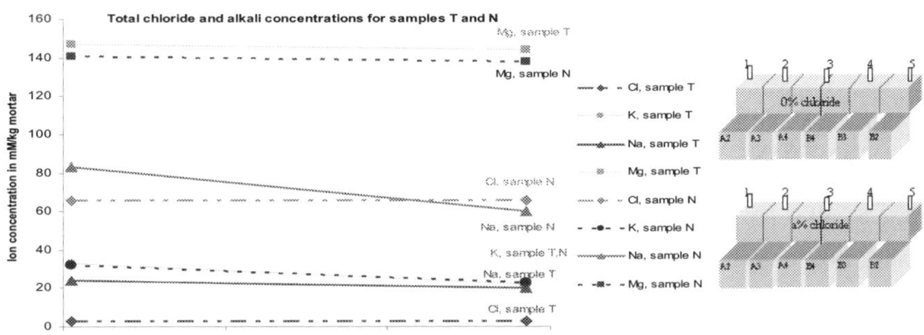

Figure 4: Chloride and alkali concentrations in the reference samples T and N.

Fig. 3 represents the changes in alkali concentrations with section locations for the TN samples. The concentrations of sodium ions differ significantly from sections A to sections B, as sections B were cast with 3% NaCl as mixing water. There is a trend towards decrease in Na^+ concentration from sections B to sections A. Up to 14 days of hydration, the Na^+ concentration is decreasing in B2 but increasing in sections A4 and A3. A slight shift towards lower values is recorded for all sections within 57 days hydration. These changes in concentration can be partly attributed to ion transport as the result of concentration gradient. As for K^+, almost constant concentration is recorded in all sections of the samples. However, Fig. 3 reveals a clear trend of decrease in the average potassium concentration during hydration. The same observations are recorded for K^+ in the T and N samples (without concentration gradients).

The relative decrease (compared to the respective initial values) of ion concentrations is about 3 % for Mg^{2+}, 30% for K^+, and ranges between 16% (for T samples) and 28% (for N samples) for Na^+. The global chloride concentrations in both T and N samples are constant during the period of experiments.

These observations suggest that in the 'mixed' TN samples, the change in alkali concentration is not mainly due to diffusion (as expected), but to another mechanism which takes place in the reference samples (T and N) as well. The decrease in alkali concentrations is probably a result of leaching, which is an inevitable consequence of shrinkage in the hydration process. As the specimens were kept in almost 100 % humidity environment, there was always a renewable water film on the mortar surfaces; hence, penetration of fresh water into the material structure is unavoidable.

3.3 Chloride concentrations vs electrical resistivity

Fig. 5 presents the evolution of mortar resistivity with respect to the aggressive ion concentration in mortars, aiming at evaluation of mortar corrosivity and the corresponding electrical properties of the materials. As previously reported [3], it is expected that changes in chemical composition, especially with the involvement of free ion movement, will cause a change in the electrical resistivity and conductivity of the mortars. However, this is not always the case, as in this study, chloride was already present in the samples in much higher concentrations than recommended thresholds [2] in terms of corrosion risk. For section A4 (Figs. 5a and 5b), there is a shift towards higher values of both total and free chloride concentrations (the total chloride more significantly) after 14 days hydration, while the resistivity values are gradually increasing during hydration. Figs. 5c and 5d represent an opposite trend (decrease in chloride concentrations) in section B4 which is attributed to diffusion from part B to part A of the prisms. It can be concluded that, in addition to external factors such as RH and curing temperature, the electrical resistivity is mainly dependent on pore structure and microstructural changes in the materials. When NaCl is admixed into the specimens, the following chemical reaction will lead to Friedel's salt formation:

$$2NaCl(aq)+3CaO.Al_2O_3.6H_2O(s)+Ca(OH)_2(s)+4H_2O \rightarrow 3CaO.Al_2O_3.CaCl_2.10H_2O(s)+2NaOH(aq) \quad (1)$$

Hence, the release of NaOH will lead to increase in the pH of the pore solution. The increasing pH will accelerate hydration process and consequently, modify the pore structure towards a finer pore size distribution. The pore structure will be investigated by Scanning Electron Microscopy techniques.

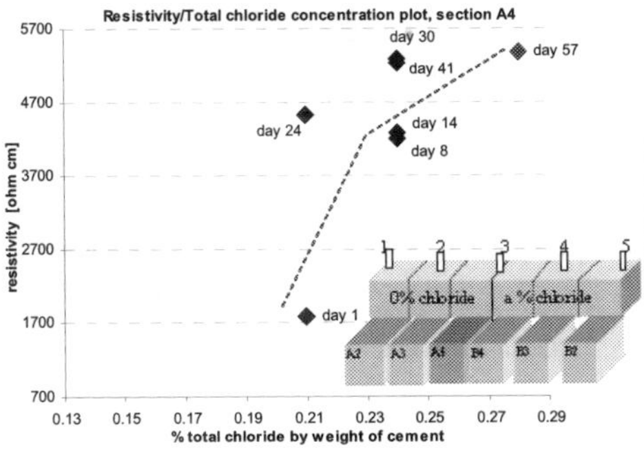

(a) Total Cl⁻ vs resistivity, section A4

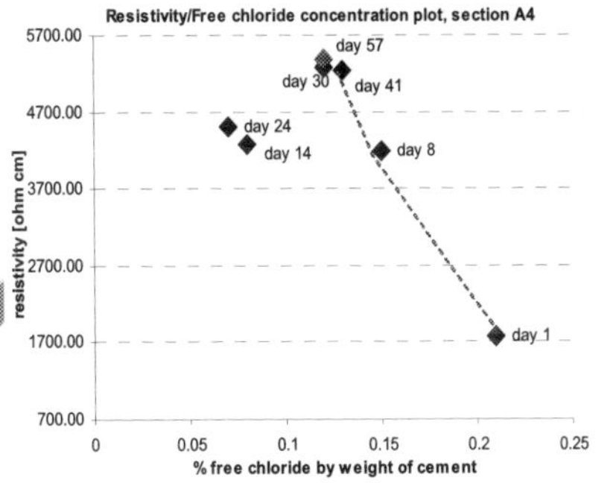

(b) Free Cl⁻ vs resistivity, section A4

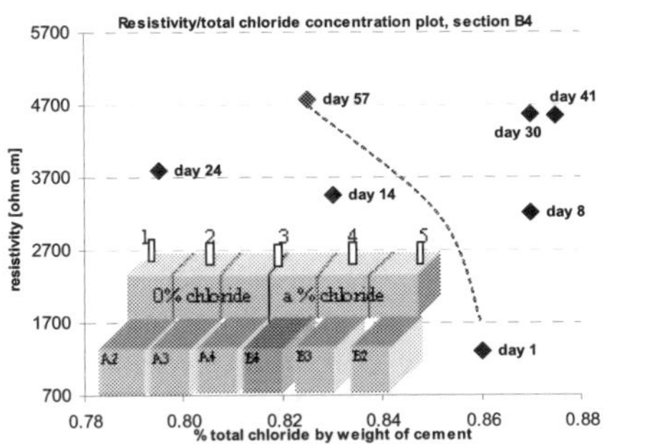

(c) Total Cl⁻ vs resistivity, section B4

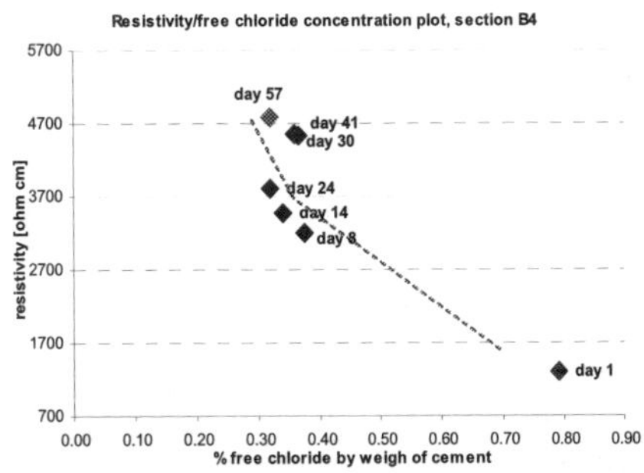

(d) Free Cl⁻ vs resistivity, section B4

Figure 5: Influences of chloride concentrations on mortar resistivity.

3.4 Determination of the pore structure parameters

The microstructure and morphology changes of the specimens were visualised and investigated using Environmental Scanning Electron Microscope (ESEM), equipped with Energy Dispersive X-ray analyser (EDAX).

3.4.1 Morphological changes

The ESEM image analysis revealed different morphology of the C-S-H in the T and N samples (Fig. 6). The differences are due to the additions of NaCl and the chloride binding, which produce different structure and phases of the C-S-H gel.

Figs. 6-8 clearly visualise the differences between T and N samples in morphology of the C-S-H hydration products, which also lead to differences in the pore structure. Referring to the differences of C-S-H morphology in Fig. 6 for T and N samples, it seems that the C-S-H gel in the N samples yields denser structure compared to the T sample. Different phases of the C-S-H in the N sample are revealed in Fig. 8, with their corresponding spectrums included.

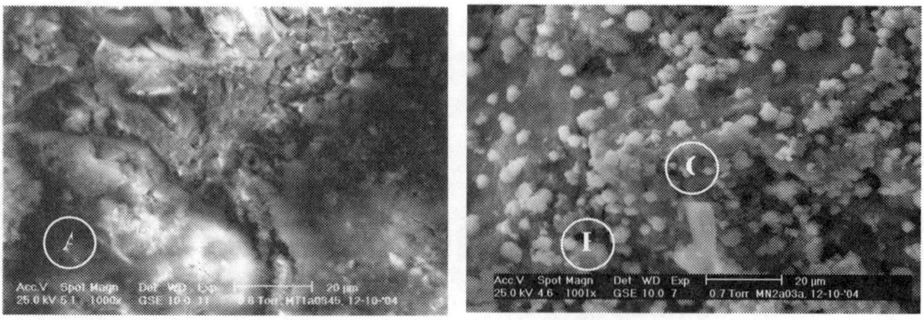

Figure 6: Overview of T (left) and N (right).

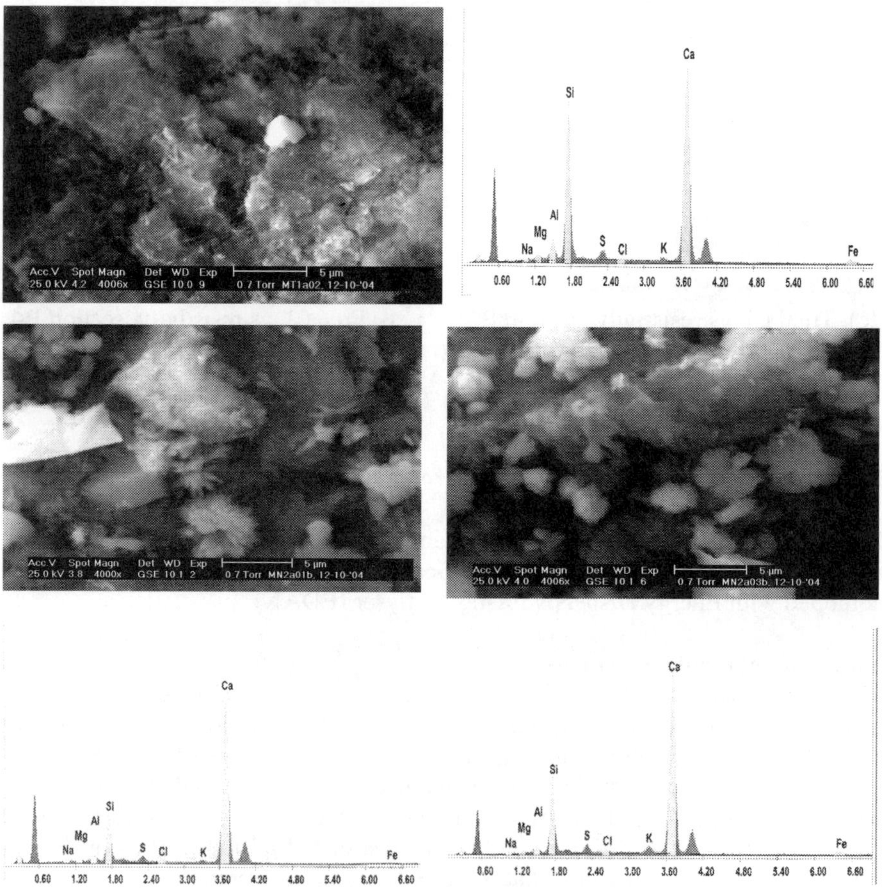

Figure 7: Local area A in Fig. 6 (left) and corresponding spectrum (right).

www.witpress.com, ISSN 1743-3533 (on-line)

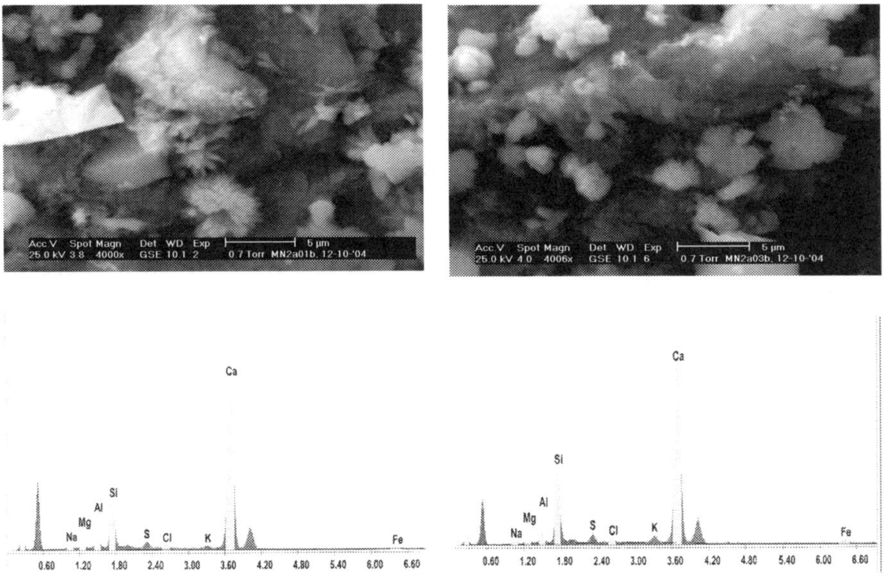

Figure 8: Local areas B and C (in Fig. 6) revealing different phases of C-S-H in sample N.

3.4.2 Backscattered electron images

Backscattered electron (BSE) images of random locations on the polished section were employed for each group of mortar specimens (Fig. 9). The physical size of the reference region of each image is 226 µm in length and 154 µm in width (712x484 pixels). The resolution in this study is 0.32 µm/pixel. The original BSE image was segmented by applying a grey-level threshold to create a binary image reflecting the pore phase. The binary image was then subjected to an image processing software for quantitative evaluation of the pore space.

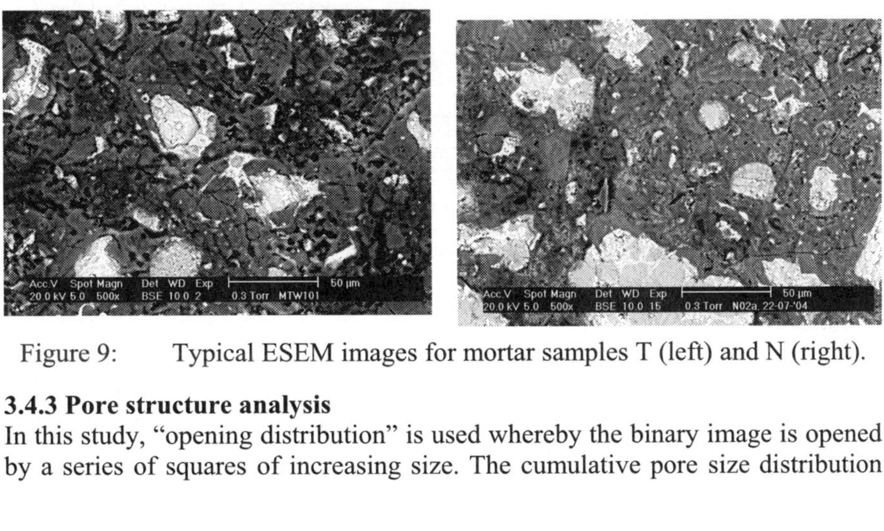

Figure 9: Typical ESEM images for mortar samples T (left) and N (right).

3.4.3 Pore structure analysis

In this study, "opening distribution" is used whereby the binary image is opened by a series of squares of increasing size. The cumulative pore size distribution curve is obtained by plotting the pore area fraction after an opening operation versus linear dimension of the structuring element. This gives a type of "size" classification in the case of an interconnected structure like pore space in mortar. The critical pore size l_c can be conceived as the diameter of the pore that completes the first interconnected pore pathway in a network developed by a procedure of sequentially adding pores of diminishing size to this network. It is generally accepted that the smaller the critical pore size, the finer the pore structure. The critical pore size is a unique transport length scale of major significance for permeability properties. The critical pore size l_c can be associated with the inflection point of cumulative pore size distribution curve. For details of opening distribution technique, see [4].

As to the morphological aspect of pore structure, specific surface area S_V is a widely used parameter. Another characterizing parameter, the so-called pore distribution density (PDD), is also measured for the samples. The pore distribution density is a three-dimensional structural parameter derived from the skeleton length of pore features observed on two-dimensional section images. This parameter is found to have higher degree of correlation with water permeability of cement pastes than the conventional parameters used for characterizing pore structure (such as porosity, critical pore size). This higher correlation can be attributed to the fact that this new parameter PDD contains information on both pore size and connectivity of pore space. The derived structural information allows prediction of the intrinsic permeability k of the mortars on the basis of empirical relationships between pore structure and permeability. For this purpose, an upgraded version of the Katz-Thompson equation is employed which incorporates the new parameter PDD. The results of structural analysis and permeability predictions are summarized in Table 1 for the two groups of mortar samples.

Table 1: Structural information derived from SEM images of specimens at 60 days of hydration.

Parameters	T (using tap water)	N (using NaCl)
Porosity (%)	30.46 (11.47%)	13.54 (12.60%)
Critical pore size l_c (μm)	1.585	0.951
Specific surface area S_V (μm^{-1})	1.088 (15.13%)	0.410 (5.64%)
Pore distribution density (-)	1.454	0.199
Intrinsic permeability k (m^2)	$5.77x10^{-17}$	$1.04x10^{-19}$

Note: The values indicated in brackets are coefficients of variation for the measured parameter.

4 Conclusions

The experiments reveal minor influences of changes in chemical composition (due to ion movement) on mortar resistivity during the period of testing. The mortar resistivity is mainly dependent on the microstructural changes due to hydration and chemical binding for the samples with aggressive ion content.

Further investigation is necessary about the influences of microstructural changes on electrical properties of the materials. Monitoring electrical resistivity is a simple technique for corrosion evaluation and prediction but it alone could not be used as a reliable technique. As in this study with high RH (>80 %) and significant aggressive ion concentrations (free and total chloride concentrations in the range of 0-0.40% and 0.04-0.88% by cement weight, respectively), there is a deviation from the rule of thumb that 'the higher the resistivity, the lower the corrosion probability'. The mortars with different chloride concentrations present similar electrical resistivity. Electrical resistivity does not significantly depend on chloride ion concentration, i.e., significant chloride ion concentration does not necessarily mean lower resistivity. Electrical resistivity is determined by both the chemistry of the pore solution and the pore structure, the latter being dependent on chemical and morphological changes under the influences of aggressive ions.

References

[1] Monfore G.E., The electrical resistivity of concrete. *Journal of the PCA Research and Development Laboratories*, pp. 35-48, 1968.

[2] Gaal, G.C.M., *Prediction of deterioration of concrete bridges*, Ph.D. Thesis. Delft: Delft University Press, 2004.

[3] Polder, R.B, Test methods for on-site measurement of resistivity of concrete - RILEM TC-154 technical recommendation. *Construction and building materials* **15**, pp. 125-131, 2001.

[4] Hu, J. & Stroeven, P., Application of image analysis to assessing critical pore size for permeability prediction on cement paste. *Image Analysis & Stereology* **22 (2)**, pp. 97-103, 2003.

Modelling the localized corrosion effects experienced by zinc–4.5wt.% aluminium steel coatings in 5% NaCl solution

N. C. Barnard[1], S. G. R. Brown[2] & H. N. McMurray[2]
[1]*Engineering Doctorate Centre, Wales, University of Wales Swansea, UK*
[2]*Materials Research Centre, School of Engineering, University of Wales Swansea, UK*

Abstract

An attempt is made to quantitatively represent the localized corrosion effects experienced by zinc and zinc-aluminium coatings in contact with aqueous chloride electrolytes. The model is intended to operate at the micro-mesoscopic scale to include features of metallurgical interest associated with these coating alloys. Using the finite difference method, field based simulations estimate the location and intensity of localized phenomena brought about by predicted dissolution events occurring at the surface of an alloy coating. Perturbations in the ion concentration are evaluated by the model to determine potential and current flux in solution. Differing coating thicknesses are considered in addition to differing microstructures resulting from production conditions. From these, normalized current density fields predicted in the electrolyte present are evaluated.
Keywords: localized corrosion, modelling, zinc aluminium steel coatings, finite difference, microstructure.

1 Introduction

There has always been an interest in protecting steel, one of the most prominent engineering materials in the construction industry against the detrimental effects of corrosion. Corrosion effects can be seen to alter material properties, and with the associated economic impact of corrosion, it is often desirable to limit the opportunity and influence of this physicochemical process on the world's durable goods and infrastructure. Metallic coatings are used in the protection of

WIT Transactions on Engineering Sciences, Vol 48, © 2005 WIT Press
www.witpress.com, ISSN 1743-3533 (on-line)

iron and steel, traditionally applied by hot-dip methods. The capacity of zinc to afford barrier and cathodic protection as a steel coating has seen the widespread use of galvanized products for many years [1]. Within the construction industry demand has increased for the improvement of corrosion resistance of organically coated steels, typically used in cladding and roofing of buildings. Significant current interest exists in the development of galvanizing coatings containing ~4.5wt-%Al. Coating alloys of this type have been seen to exhibit greater ductility and superior corrosion resistance than conventional galvanizing alloy systems (galvanized steels suffer larger Zn dissolution rates [2]) some of which include small aluminium additions in order to prevent large scale intermetallic formation during fabrication [1, 2, 3, 4].

The 4.5 wt-% Al zinc alloy, often known as Galfan, is commonly applied to steel and used as the substrate for organically coated steel products, used as profiled cladding in building solutions. In such applications the product is susceptible to localized degradation at cut-edges leading to delamination of organic coating and degradation of metallic coating remaining exposed. Galfan owes its apparent improved corrosion performance over conventional systems to its microstructure. The eutectic point in the zinc–aluminium alloy system exists around 5 wt-%Al. The proeutectic 4.5 wt-% Al composition leads to a microstructure consisting of Zn crystals (where Al is almost never solid-soluble) containing about 1% Al surrounded by a lamellar eutectic of zinc containing ~5% aluminium [3, 5]. The zinc–aluminium eutectic retains a comparable cathodic protection as that offered by pure zinc coatings and also has an improved resistance to corrosion owing to the development of an adherent aluminium oxide layer over exposed areas. However, zinc alloy coatings of this type do suffer from non-uniform attack due to changes in the microstructure brought about by impurities and additions to the hot-dip bath. The main failure mechanism of zinc–aluminium galvanized steels is the removal of zinc from the coating layer via anodic dissolution, eqn. (1), the progression of which is electrochemically supported by the cathodic reduction of oxygen, eqn. (2).

$$Zn_{(s)} \rightarrow Zn^{2+}_{(aq)} + 2e^{-} \tag{1}$$

$$O_{2(g)} + 2H_2O_{(l)} + 4e^{-} \rightarrow 4OH^{-}_{(aq)} \tag{2}$$

Material losses from the cut-edge situation can lead to substantial undermining of the organic layer resulting in further degradation.

An attempt is made here to predict the localized corrosion effects experienced by these zinc alloy coatings in the presence of an aggressive electrolyte, 5% NaCl solution. The model described has been developed primarily from a metallurgical standpoint rather than an electrochemical one [6]. In order to reflect the influence of microstructure in these coatings (which varies with production conditions) the model was developed at the micro-mesoscopic scale rather than the molecular level. This means that features of metallurgical interest are represented and the associated changes in corrosion performance successfully predicted. All simulations depicted here involve contact with 5% NaCl pH neutral media and are isothermal at 298K such that comparison can be made to accelerated corrosion tests performed via the scanning vibrating electrode technique (SVET) under the same conditions.

2 Modelling procedure

An orthogonal structured computational grid is used by the model to represent both the solid and liquid within each simulation. Differential equations employed by the model are approximated via straightforward finite difference methods allowing for variable nodal spacings in the x, y and z directions. Each cell in the grid assumes an integer value identifying the site as either electrolyte or solid. Electrolyte sites then also possess detail of concentration (c), electrical potential (ϕ), current density vector (i_x,i_y,i_z), volume fraction of the cell occupied by corrosion product (p) and the volume fraction of the cell that has been corroded away. Solid sites contain information of composition and volume fraction corroded (f). The discretized representation of the coating layer can be seen in figure 1(a) which is an approximation of a Zn–4.5 wt-% Al steel coating based on observations made by Elvins et al. [3]. Figure 1(a) shows the formation of primary zinc crystals within eutectic and this is the setup which is used in simulating the surface corrosion performance of 4.5 wt-% Al zinc alloy coatings. The arrangement of the simulated cut-edge case is shown in figure 1(b). The coating layer (darker region) is constructed in the same way as that for the surface case. The presence of a steel substrate (central lighter area) is included in addition to an inert polymeric material (white) which restricts contact between the coating 'surface' and electrolyte.

2.1 Corrosion

Depicting the localized corrosion of zinc alloy steel coatings with a physically based numerical model required simplifying assumptions to be made. The first order model deals with the evolution and movement of a single ionic concentration, increases of which are brought about by anodic dissolution, in the case of zinc eqn. (1). The model assumes that the Zn^{2+} rapidly undergoes hydrolysis to produce solid zinc hydroxide via

$$Zn^{2+}_{(aq)} + H_2O_{(l)} \rightarrow Zn(OH)^{+}_{(aq)} + H^{+} \tag{3}$$

$$Zn(OH)^{+}_{(aq)} + H_2O_{(l)} \rightarrow Zn(OH)_{2(s)} + H^{+} \tag{4}$$

Thus the models track the effective free proton concentration arising from the dissolution of zinc and aluminium throughout the corrosion process. Direct estimation of the overall corrosion effects experienced considering only one electroactive species distinguishes the model from those that deal with individual species [7, 8]. Distribution of Zn^{2+} and pH across a galvanic couple can be seen to support this approach due to the pivotal role of corrosion products in the model [9, 10].

www.witpress.com, ISSN 1743-3533 (on-line)

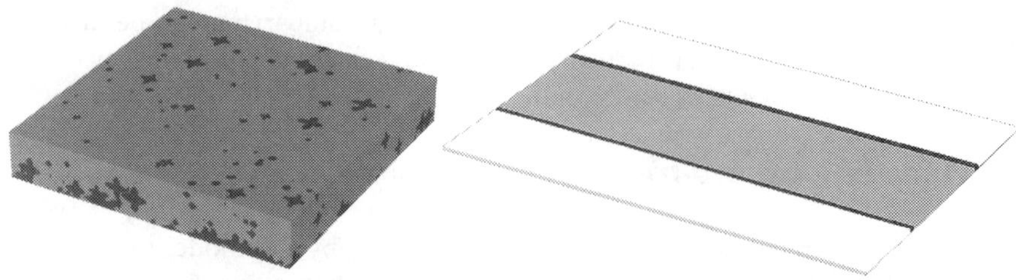

Figure 1: Representation of solid constituents (a) Galfan coating and (b) cut-edge Galfan coated steel.

2.2 Electrical potential

Interface sites identified in the model as liquid sites adjacent to one or more solid sites are the subject of a calculation to determine the reversible potential, ϕ_0 (V) including perturbations in concentration, c ([$H^+_{(aq)}$]), deviating from the initial concentration, c_0, using eqn. (5). H^+ is frequently potential determining for metal surfaces in contact with aqueous solutions (due to the hydroxylated nature of the interface metal (oxide) film).

$$\phi_0 = E_0 - \frac{RT}{nF}\ln\left(\frac{c}{c_0}\right) \tag{5}$$

where R is the universal gas constant, F is the Faraday constant, T is the temperature, E_0 is the standard electrode potential (V) determined from a weighted average of all solid cells in the local Moore neighbourhood. These potentials are used as Dirichlet boundary condition to determine the electrical potential in the remaining electrolyte cells using

$$\nabla.(-\kappa\nabla\phi) = 0 \tag{6}$$

where κ is the conductivity ($\Omega^{-1}m^{-1}$).

The rate at which dissolution occurs at interface sites is related to the potentials experienced at those positions. Kinetics of dissolution for phases present at the solid–liquid interface are approximated from experimentation using the rotating disc electrode (RDE) technique, detailing the Tafel relationship for pure zinc and commercially pure aluminium samples. The rate of dissolution, a function of current density, is derived from the potentials associated with the composition of a cell, and consequently from eqn. (5) in the case of this model, $j \propto [H^+]$.

2.3 Localized dissolution

The amount by which each corroding metal/electrolyte surface cell corrodes is calculated from the local current density and the volume fraction of material dissolved in each cell is updated. Once a cell is completely dissolved it is redefined as an electrolyte cell. Therefore over a time increment the total increase in concentration in the evolving system is quantified via

$$C_A = \sum_{i=1}^{n} \delta f_i c_a \Delta x_i \Delta y_i \Delta z_i \tag{7}$$

Where n is the number of sites corroding, f_i is the volume fraction of a cell transforming, and c_a is a constant volume concentration contribution term.

2.4 Movement of electroactive species

The movement of charged particles in solution, which are evolved during localized corrosion, is modelled by laws governing convection, diffusion and the transport of ions through an electrical field. Diffusion through 5% NaCl solution (negligible convection) over a time increment due to ordinary diffusion and migration effects is modelled by eqn. (8).

$$\frac{\partial c}{\partial t} = -\nabla.(D\nabla c) + zF\nabla.(uc\nabla\phi) \tag{8}$$

Where D is the diffusion coefficient at 298K (m^2s^{-1}), z is the unit charge and u is the ionic mobility ($m^2/(Vs)$).

2.5 Cathodic control

Left unregulated the localized corrosion effects experienced predicted by the model would propagate unopposed. This would not be realistic since the cathodic reaction is the rate determining process. To ensure electroneutrality in the model there is a counterbalancing procedure to eqn. (7), whereby an equal and opposite concentration decrease is forced at all metal/electrolyte interface sites,

$$C_C = \frac{C_A}{V_{SE}} \tag{9}$$

where V_{SE} is the total volume of all electrolyte cells at the solid-electrolyte interface able to accept concentration. There is also a lower limit placed on cell concentration such that sites acting cathodically approaching this limit will cease to lower concentration, lowering the overall corrosion rate.

Another mechanism regulating the localized corrosion effects in the model are the actions of corrosion products. Deposited at surface sites within pre-determined concentration bands these sites hinder the diffusion of ionic species through them by reducing the diffusion coefficient according to the volume of the cell occupied by corrosion product (p_i),

$$D_i = D(1 - p_i)^2 \tag{10}$$

If p_i reaches unity then that site acts as a complete barrier to diffusion. At each time interval the volume of corrosion product deposited at the surface is 5% of the volume of the metal dissolved and all sites containing corrosion product have their fraction, p_i, reduced by an arbitrary amount to simulate gradual dissolution of corrosion product.

2.6 Current density

The current densities presented in section 3 are normalised vertical current densities calculated via numerical differentiation of the electrical fields generated using eqn. (11). This post-processing enables comparison of the predictions made by the computer model to experimentally derived data, e.g. SVET data, and the form of observed current density fields can be seen to match numerical solution. [11, 12]

$$i_y = -\frac{1}{\beta}\frac{\delta\phi}{\delta y} \tag{11}$$

where β is the electrical resistivity (Ωm) of 5% NaCl solution.

3 Results and discussion

The voltage fields shown in figure 2 correspond to the voltage fields throughout 5% NaCl solution in contact with the simulated cases represented in figure 1. In the case of surface corrosion, figure 2(a), the anodic potentials correspond to the presence of primary zinc dendritic regions present at the surface, and cathodic voltages occur at spatially discrete areas corresponding to eutectic material. Figure 2(b), cut-edge corrosion simulation, the anodic potentials arise almost exclusively on sites occupied by Galfan coating, only in this case the majority of the cathodic potentials are predicted on the steel substrate. In both cases the localized effects can be seen to have far reaching effects into the electrolyte.

Figure 3 shows the predicted normalised current density fields 100μm above the corroding surface of simulated Galfan coatings, identical in properties and mesh size, only representing different areas of exposed coating. In both cases the form is what would be expected and is similar to those predicted by SVET experimentation. Current densities in figure 3(b) can be seen to closely correlate both in frequency and relative intensities to those exhibited in an equivalent area in figure 3(a) suggesting that in the case of modelling the surface corrosion of Galfan, predictions are independent of grid size [6].

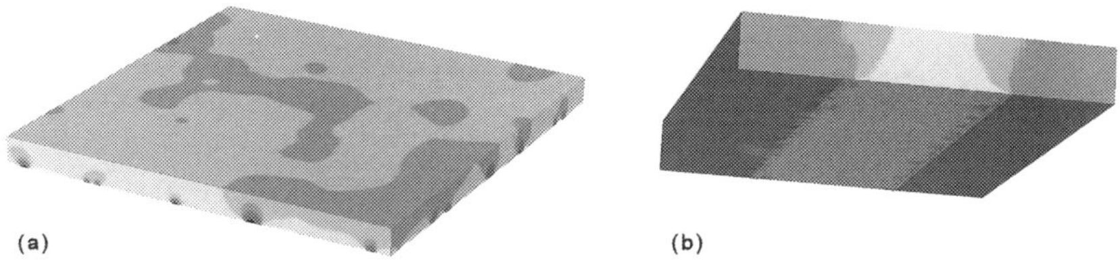

Figure 2: Electrical potential fields in electrolyte above corroding (a) Galfan surface and (b) cut-edge Galfan coated steel samples after 12 hours.

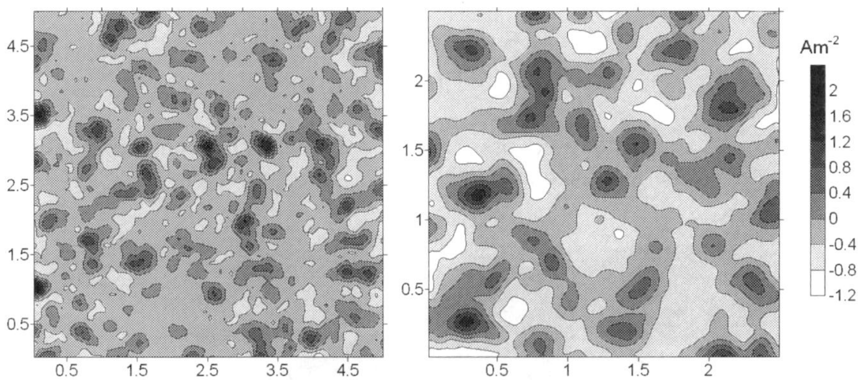

Figure 3: Current density field above corroding Galfan surface (a) 5mm×5mm and (b) 2.5mm×2.5mm after 18 hours immersion.

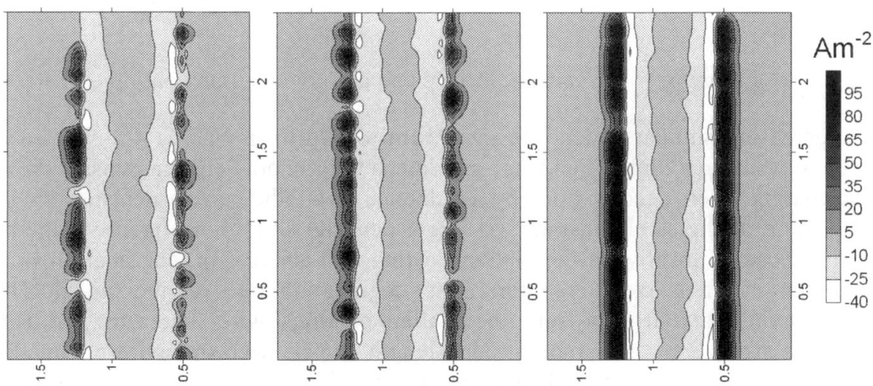

Figure 4: Current density exhibited by simulated Galfan coated steel cut-edge samples reflecting (a) fast cooling, (b) moderate cooling and (c) slow cooling of the hot-dip coating after 12 hours immersion. (Dimensions in mm).

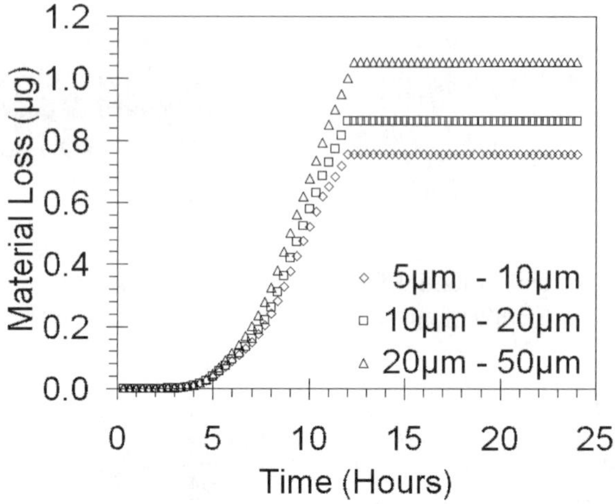

Figure 5: Variation of zinc loss from cut-edge over time comparing different primary phase dimensions.

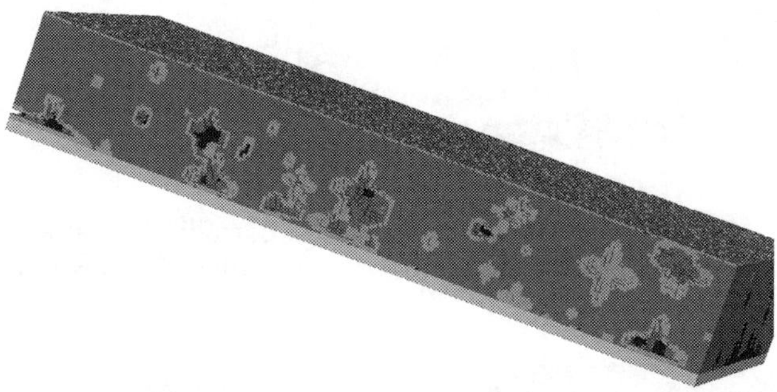

Figure 6: Exposed Surface and cut-edge Galfan coating.

Increasing simulated dendrite size (5µm to 50µm) in figures 4(a-c), can be seen to result in a higher incidence and intensity of anodic current density above the coating layers accompanied by a reduction in anodic current densities above the centre of the steel substrate. At larger dendrite sizes more intense cathodic activity occurs locally on the Galfan coating and above adjacent steel areas, a situation which is conducive for higher degradation rates in the model. The increase in material lost from the Galfan coatings with increasing primary dendrite sizes over the same cut-edge length can be seen in figure 5. Smaller dendrites give rise to higher degradation rates early in the respective simulations. This is due to the increased number of potential local anodic sites. The small size of these dendrites means their sacrificial protection is available quickly to the steel in the simulation. When the simulated corrosion process is curtailed, lower degradation rates are exhibited than simulations with increased primary dendrite sizes.

Figure 6 shows the solid material during the latter stages of simulated localized corrosion exhibiting substantial loss of primary zinc phase. Deposition of corrosion product leads to a reduction in corrosion rate due to covering of metallurgical features more susceptible to localized anodic dissolution.

4 Conclusions

A prototype first-order model of localized corrosion on Zn-Al steel coatings has been presented. The overall behaviour of the model appears sensible and the model is in good agreement with experimentally measured SVET data. While the model is far from perfect, it does represent a step towards predictive 3D simulation of localized corrosion processes.

Acknowledgements

This work was carried out under the Engineering Doctorate Scheme, the help and support of Corus Group and the funding of EPSRC are gratefully acknowledged.

References

[1] McMurray, H.N., Parry, G. & Jeffs, B.D., Corrosion resistance of Zn–Al alloy coated steels investigated using electrochemical impedance spectroscopy. *Iron and Steelmaking*, **25(3)**, pp. 210–215, 1998.

[2] Li, Y., Corrosion behaviour of hot dip zinc and zinc-aluminium coatings on steel in seawater. *Bulletin of Materials Science*, **24(4)**, pp. 355–360, 2001.

[3] Elvins, J., Spittle, J.A. & Worsley, D.A., Relationship between microstructure and corrosion resistance in Zn–Al alloy coated galvanised steels. *Corrosion Engineering, Science and Technology*, **38(3)**, pp. 197–204, 2003.

[4] Shih, H.C., Hsu, J.W., Sun, C.M. & Chung, S.C., The lifetime assessment of hot-dip 5% Al-Zn coatings in chloride environments. *Surface and Coatings Technology*, **150**, pp. 70–75, 2002.

[5] Hada, T., Production technologies of hot-dip zinc-aluminium alloy coated steel sheet. *Proceedings of the 4th International Conference on 'Zinc and Zinc Alloy Coated Steel Sheet'* (Galvatech'98), Chiba, Japan, Iron and Steel Institute of Japan, pp. 108–114, 1998.

[6] Brown, S.G.R., Barnard, N.C. & McMurray, H.N., 3-dimensional modelling of localized corrosion effects in zinc and zinc alloy steel coatings. *Proceedings of the 3rd International Conference on 'Computational modeling and simulation of material'* (CIMTEC 2004), Acireale(CT), Italy, pp. 393–400, 2004.

[7] Johnsen, T., Jøssang, A. & Meakin, P., An experimental study of the quasitwo- dimensional corrosion of aluminum foils and a comparison with

two dimensional computer simulations. *Physica A*, **242**, pp. 356–376, 1997.

[8] Cordoba-Torres, P., Nogueira, R., de Miranda, L., Brenig, L., Wallenborn, J. & Fairen, V., Cellular automaton simulation of a simple corrosion mechanism: mesoscopic heterogeneity versus macroscopic homogeneity. *Electrochemica Acta*, **46**, pp. 2975–2989, 2001.

[9] Tada, E., Satoh, S. & Kaneko, H., Distribution of Zn^{2+} during galvanic corrosion of a Zn/steel couple. *Electrochemica Acta*, **49**, pp. 2279–2285, 2004.

[10] Tada, E., Sugawara, K. & Kaneko, H., Distribution of pH during galvanic corrosion of a Zn/steel couple. *Electrochemica Acta*, **49**, pp. 1019–1026, 2003.

[11] Isaacs, H., The effect of height on the current distribution measured with a vibrating electrode probe. *Journal of Electrochemical Society*, **138(3)**, pp. 722–728, 1991.

[12] Livingstone-Bridge, D., Myland, J.C. & Oldham, K.B., A model of ionic current densities in the vicinity of a corroding disk shaped region. *Electrochemistry Communications*, **3**, pp. 384–389, 2001.

Numerical simulation of metal corrosion with cluster formation

A. Taleb[1], J. Stafiej[2] & J. P. Badiali[1]
[1]*Laboratoire d'Electrochimie et de Chimie Analytique, ENSCP et Université P. et M. Curie, France*
[2]*Institute of Physical Chemistry, Polish Academy of Science, Poland*

Abstract

Using a cellular automaton model we describe the corrosion of a piece of metal in contact with an aggressive medium. The metal considered contains some defects that are distributed on some lines crossing the overall metal. Corrosion is considered in a particular environment where no oxide film is formed on the metal surface and where the diffusion is not a determining process. The main parameters of the model are the number of line-defects and the corrosion probability of the metal surface site. This probability is larger when a site is located on a line-defect than otherwise. Varying the model parameters we investigate the roughness of the metal surface and the formation of clusters that are detached from the corrosion front. We show that a complicated surface structure may appear with a high roughness if the density of defects is not too high. A relation between the cluster concentration and the surface roughness is established.
Keywords: numerical simulation, cellular automaton model, metal corrosion, crystalline defects, clusters formation.

1 Introduction

For many metal/ environment interfaces systems there is a clear evidence that corrosion is associated with defects, inclusions, phase boundaries, etc)Evans [1] Janik-Czachor [2,3], Richardson and Wood [4], Sedriks [5]). In some cases it was observed that the effective dissolution rate is larger than the one expected for pure electrochemical processes suggesting that another dissolution process parallel to the electrochemical ones occurs simultaneously (Marsh and Schaschal

[6], Garreau and Bonora [7], Bech et al. [8]). In some conditions related to the surface roughness an additional mechanical process may take place: pieces of metal detached from the surface appear as clusters diffusing in the solution. This mechanical disintegration of the metal is known as chunk-effect (Marsh and Schaschal [6], Garreau and Bonora [7], Bech et al. [8]). The main part of this paper is devoted to an investigation of this coupling between electrochemical and mechanical processes.

In the theoretical description of corrosion processes, we have to deal with a large number of chemical species, a combination of many processes (chemical, electrochemical, mechanical,) and we have to perform a multi-scale analysis both in size and time. However, to understand how a few numbers of simple processes generally involved in corrosion processes behave may provide valuable insights into the analysis of experimental results and may stimulate new research directions. Of course, for actual systems we are not free to select a few numbers of such processes but this is easy in computer simulations.

The main goal of this paper is not to explain the properties of a given interface but to create a simple model from which we may analyze how the combination of some simple processes may determine the surface properties and the formation of clusters.

This paper is organised as follows. In section 2, we describe the physical model. In section 3 we describe the automata cellular model. In Section 4 we consider two limiting cases of our model. In Section 5 we present the simulation results and a discussion of them.

2 Physical picture of the system model

Our approach corresponds to a description at mesoscopic scale in which we perform implicitly a coarse graining in space and a time average on processes developing at a microscopic scale, these processes generate the random aspects in the model.

In the system model we investigate the metal dissolution results from the combination of electrochemical (corrosion) and mechanical (cluster formation) processes. We consider that the dissolution of corrosion products is rapid in comparison with other processes and there is no film formation on the metal surface. The corrosion products are assumed to diffuse rapidly in the solution (i.e. the corrosion products disappear immediately after their formation). Finally, we assume that the metal is a real system in which it exists, a given number of crystalline defects on which the corrosion rate is enhanced.

Hereafter, based on these considerations, we construct a simple cellular automata model for investigating the structure of the corrosion front and the formation of clusters.

3 The automata cellular model

As it is traditional in simulations of growth on the surface (Nainville et al. [9], Taleb et al. [10], Cordoba et al. [11], Malki et al. [12], Kortlöke [13]), the simulations are performed on a two dimensional square lattice. In this infinite lattice, we isolate a slice in which there are N_x sites in the Ox direction and N_y in the normal direction Oy. Each site i is referenced by two integer numbers n_i and m_i which are its coordinates in the Ox and Oy direction respectively; n_i runs from 0 to N_x and m_i from 0 to N_y. We assume that the part of the lattices adjacent to this slice is such that we have periodic boundary conditions.

Each lattice site is occupied by a given species involved in the corrosion process. At the beginning of the calculation, whatever the value of n_i the site corresponding to $m_i \geq m_0$ are occupied by the species S (solvent or aggressive species contained in the solution), while the metal is associated with $m_i < m_0$.
To mimic the presence of defects such as straight dislocations (edge or screw) in the metal we introduce a number D of columns equally separated in which the metal atoms have special properties with respect to the corrosion. This is a very crude model, it should be more realistic to consider that a line-defect has a given thickness, that it is not a straight line and that the lines are randomly distributed in the metal. The model that we consider contains the minimum number of parameters and, as we shall see, it is sufficient to understand some basic effects initiated by the presence of defects.

During the simulations the metal sites isolated from the solvent are considered as bulk metal sites, if there are located on a line-defect they are noted M_f and M otherwise. The metal sites in contact with the solution are named corrosion sites and noted M_{fs} and M_s depending whether they are or not located on a line-defect.

Now we have to decide how to change the nature of the species on the lattice. In our model, we have four sites M, M_s, M_f, M_{fs} that refer to metal species but occupying different positions as described above. In our mesoscopic description each of these sites contains a given number, N_m, of metal atoms and the elementary act in our corrosion process is the destruction of a corrosion site (M_s or M_{fs}) i.e. the collective detachment of N_m atoms from the substrate. We assume that these atoms react with the species S to give corrosion products P. Note that the disappearing of a corrosion site may generate new surface sites.

This is an extremely simple picture compared to what happens in real systems in which, for instance, several chemical steps may determine the solvent reduction, the pH of the solution may play an important role, not only one solution species dictates the behavior of the corrosion process. Moreover the diffusion of P species may block temporarily corrosion sites modifying the initial corrosion rate as already investigated for D=0 (Taleb et al. [10], Hasseb et al. [14]). Nevertheless the simple model we consider can be useful to understand basic facts in corrosion.

In the realisation of this model at each time step all the surface sites are chosen in a random order and we try to transform them with a given probability: p for M_s and p_f for M_{fs}. These probabilities mimic corrosion rates that we can calculate, at least in principle, by considering the interaction between the metal atoms located on a corrosion site and one S species. These probabilities have a chemical and/or an electrochemical nature. The probability p may be associated with a generalized corrosion process and p_f is chosen larger than p assuming that the corrosion is enhanced on a line-defect. From the simulations we may extract two quantities. The mean position of the corrosion front is defined according to

$$\left\langle h_{corrosion}(N_t)\right\rangle = \frac{1}{N_x}\sum h_{corrosion}(i) \tag{1}$$

In which $h_{corrosion}(i)$ is for the column i the highest value of j for which there is an M_s or M_{fs} sites. The roughness of the corrosion front is defined according to

$$\sigma_{corrosion}(N_t) = \left[\frac{1}{N_x}\sum_{i=1,N_x}\left(h_{corrosion}(i) - \left\langle h_{corrosion}(N_t)\right\rangle\right)^2\right]^{1/2} \tag{2}$$

Finally our automata cellular model contains three parameters: the two corrosion probabilities p and p_f and the number, D, of line-defect in the simulation.

Before performing the simulations we consider two limiting cases of this model.

4 Two limiting cases of the model

4.1 Eden-like model

If we take $p = p_f$, the defects do not play any particular role and the algorithm is then very similar to the one of the Eden model (Eden [15]) used for investigating growth processes on a surface. In that case, corrosion looks like a growth in the metal side. However we have to point one major difference. In a growth process we increase the volume formed on the surface to each step but in our corrosion process some part of the corrosion front can be detached from it generating some clusters that are injected into the solution (Vautrin-Ul et al [16]). Note also that a pure Eden model should correspond to $p = p_f = 1$. Despite of these differences it has been checked that the roughness of the corrosion front versus N_t verifies a power law $\sigma_{corrosion}(N_t) \approx N_t^{\alpha}$ where the magnitude of α is not far from 1/3, the value expected from the Eden model.

4.2 Pure defect model

If we take $p = 0$ and $p_f \neq 0$ the corrosion is only located on the defects and we have a very crude model of a pitting corrosion. Neglecting any diffusion process a simple mean field model can be introduced. After N_t steps the bottom of the defect is located, in average, at the position $p_f N_t$ below m_0 (here we take $m_0=0$) and the mean position of the corrosion front is simply given by $< h_{corrosion}(N_t)> = p_f N_t D/N_x$. By using eqn (2) we can see immediately that

$$\sigma_{corrosion}(N_t) = p_f N_t (D/N_x)^{1/2} (1 - D/N_x) \tag{3}$$

showing, as expected, that we are no more in the Eden-model universality class. The eqn (3) shows that $\sigma_{corrosion}(N_t)$ disappears if the density of defects (D/N_x)

goes to zero or one, as expected. Note that in the pure defect model the roughness is not connected with the random choice of the corrosion sites as in the Eden model but it exists even in a pure deterministic regime as illustrated by eqn (3).

This brief analysis of these two limiting cases shows that the simulation results, in general, will exhibit a cross over between these two models.

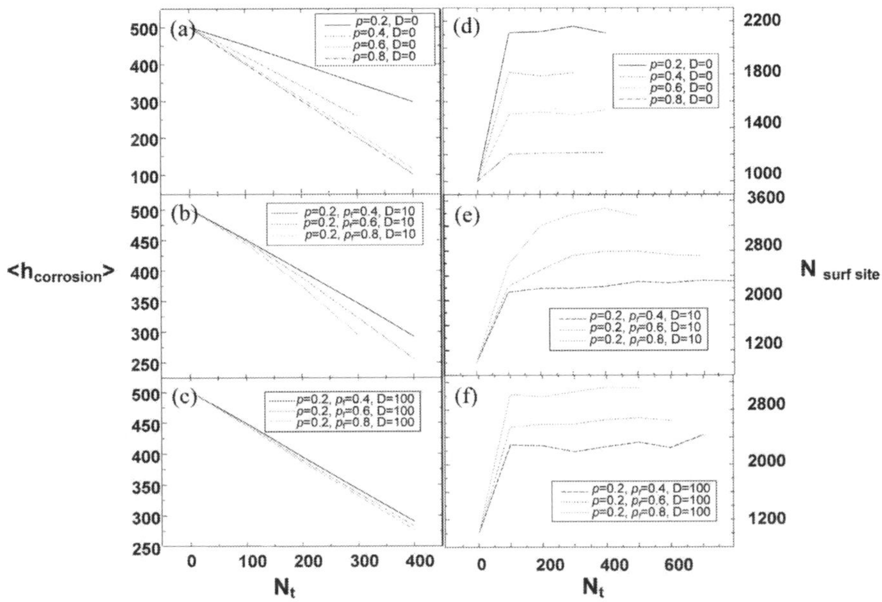

Figure 1: The evolution of the corrosion front <$h_{corrosion}$> ((a), (b), (c)) and the variation of the number of the surface sites Ms and Mfs ((d), (e), (f)), versus the simulation time step (N_t). (a), (d) the Eden like model with p=0.2, (b), (e) defect model with the number of defects D=10. (c), (f) defect model with the number of defect, D=100.

5 Discussion of the simulation results

The simulations corresponding to the Eden-like model defined in §4.1 show that the position of the corrosion front exhibits a linear dependence versus N_t as illustrated on the figure 1(a). However we observe that the results do not coincide if N_t is rescaled into N_t /p which is a rather unexpected result. This is illustrated on the figure 1(d) showing that the number of corrosion sites depends strongly on p in the explored domain of N_t. We see that lower is p larger is the number of corrosion sites and then more rough must be the surface. This is effectively observed on the figure 2(a) where we see that the roughness increases when we decrease p. This trend can be simply understood as the following: smaller is p smaller is the efficiency of the corrosion but at the same time the corroded sites are, in average, very well separated (their distance is roughly 1/p) and this process favours the formation of a large roughness. Of course higher is the surface roughness higher is the probability to form some kind of peninsula on the surface and then the detachment of these peninsula leads to the creation of clusters injected in solution. On the figure 2(d) we can see that the number of such clusters increases when we decrease p. The snapshot on figure 3(a), 3(b) shows that the surface roughness depends weakly on the simulation time step. This is a consequence of the equilibrium between the corrosion process which induces an increasing of the roughness and the clusters formation which reduces this later.

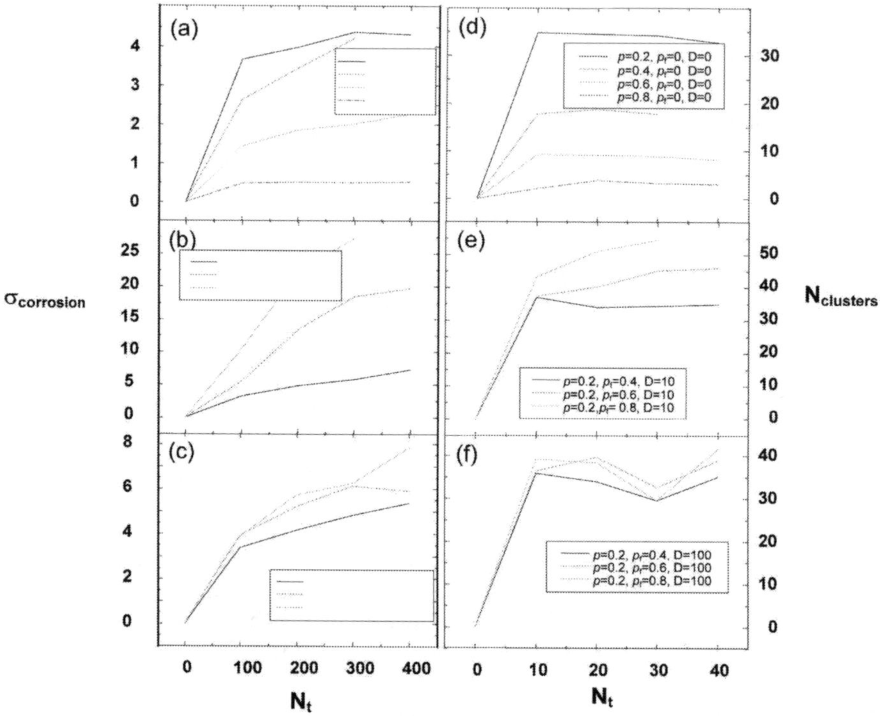

Figure 2: The roughness of the corrosion front ($\sigma_{corrosion}$) ((a), (b), (c)) and the variation of the number of cluster (N_{clusters}) ((d), (e), (f)) versus the simulation time step (Nt). - (a), (d) the Eden like model with p=0.2, - (b), (e) defect model with the number of defect D=10. - (c), (f) defect model with the number of defect, D=100.

If keeping p=0. 2 we introduce defects with the density D = 10 and increasing values of p_f, we observe on figure 1(b) that p_f increases the efficiently of the corrosion i.e. the slope of the position of the corrosion front versus N_t as expected. In parallel, the number of corrosion sites is increased, fig 1(e), and we can see on figure 2(b) that the roughness is enhanced in comparison with the case D=0, fig 2(a). For p_f =0.8 we may see that the roughness is closed from the value estimated by using eqn (3) and then we are near the pure defect model. However, even in this case the results are more than just a superposition of the two limiting models since for p = 0.2 and p_f =0.8 the number of clusters in the solution is much larger than in the case p_f =0, fig 2 (d), (e), (f).

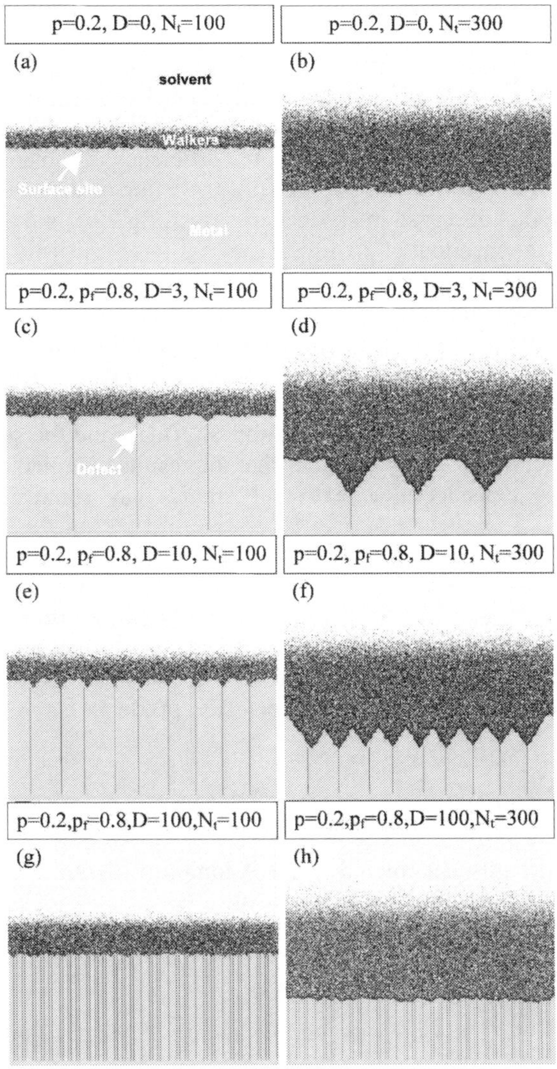

Figure 3: Snapshots taken at the N_t=100 and N_t=300 time step. The colors are ascribed to the sites as in (a) and (c). - (a), (b) The Eden like model. - (c), (d), (e), (f), (g), and (h) the defect model.

If the density of defects is increased from D=10 to D=100 the interaction between defects and the coupling between defects and ordinary surface sites (M_s) destroy the previous behaviour. The interpretation of this result is quite clear. As the ratio p_f/p increases, the probability to form pits on the surface site with defect increases.

At the beginning of the simulation (N_t=100), for D=3 and D=10, the pits grow independently, fig 3(c), 3(e). Deep pits are formed before their interaction by the overlapping of their perimeter which explains the higher roughness observed, fig 2(b). Now, by increasing the number of defect (D=100) we decrease the distance between the pits and the overlapping of their perimeter takes place already for N_t=100, fig 3(g). This explains the low surface roughness observed in this case, fig 3(g), 3(h)). With high defect number, the efficiency of the corrosion becomes weakly sensitive to the value of p_f , fig 1(c), the value of the roughness, fig 2(c), is similar to the one observed in absence of defect, fig 2(a), and the production of clusters is now independent of p_f, fig 2(f).

All the results presented above can be summarized as follows. For weak values of p and low number of defects (D) the efficiency of the corrosion is low but we may create a complicated surface structure with a high roughness and a high production of clusters i.e. a strong coupling between electrochemical and mechanical processes. In opposite, for large values of p and large concentration of line-defect there is a general cancellation of effects and the surface exhibits a rather poor structure. We may consider that the results obtained above for which the diffusion is considered as extremely rapid may mimic an atmospheric corrosion.

In order to conclude: we have investigated a simple model in which a generalized corrosion (p) takes place in a material where there is a given number D of line-defect characterized by a corrosion rate $p_f > p$. We observe a coupling between electrochemical and mechanical properties of the surface which gives a simple description of the chunk effect. In subsequent paper we will analyze how the diffusion of the corrosion products may change the result obtained above.

Acknowledgment

The authors gratefully acknowledge the Polonium program for supporting this research.

References

[1] Evans, U. R., The Corrosion and Oxidation of Metals: Scientific Principle and Practical Application, Arnold, London, 1960.
[2] Janik-Czachor, M., J. Electrochem. Soc. 128, 513, 1981.
[3] Janik-Czachor, M., Wood, G. C. and Thompsom, G. E., Br Corros. J. 15, 154, 1980.
[4] Richardson, J. A., and Wood, G. C., J. Electrochem. Soc. 120, 193, 1973.
[5] Sedriks, A. J., Int. Metals Rev. 28, 295, (1983).

[6] Marsh, G. A., and Schaschal, E., J. Electrochem. Soc. 107, 12, 960, 1960.

[7] Garreau, M., Bonora, P. L., J. Appl. Electrochem. 7, 197, 1977.

[8] Bech-Nielsen, G., De Foutenay, F., Poulsen, H., Electrochemica. Acta, 42, 1847, 1997.

[9] Nainville, I., Lemarchand, A., Badiali, J. P., Phys. Rev. E, 53, 2537, 1996.

[10] Taleb, A., Stafiej, J., Chausse, A., Messina, R., Badiali, J. P., Electroanal. Chem. 500, 554, 2001.

[11] Cordoba-Torres, P., Nogueira, R. P., de Miranda, L., Brenig, L., Wallenborn, J., Fairén, V., Electrochemica Acta, 46, 2975, 2001.

[12] Malki, M., Baroux, B., Corr. Scie. 47, 171, 2005.

[13] Kortlöke, O., J. Phys. A: Math.Gen. 31, 9185, 1998.

[14] A. S. M. A. Haseeb, A. S. M. A., Schiraldi, P. L., Bolzan, A. E., Piatti, R. C. V., Salvarezza, R. R., Arvia, A. J. J., J. Electroanal. Chem 500, 543, 2001.

[15] Eden, M., Proceedings of the fourth Berkley Symposium on Mathematical Statistics an Probability; University of California Press 1: Berkeley, CA, Vol4, 223, 1960.

[16] Vautrin-Ul, C., Chaussé, A., Stafiej, J., Badiali, J. B., Cond. Matter. Phys., In press, 2005.

Temperature and heat transfer profiles of a fluid film squeezed between two permeable isothermal discs rotating within a magnetic field

B. Bhatt & R. Mohais
The University of the West Indies, St. Augustine, Trinidad

Abstract

We consider the case of an electrically conducting fluid squeezed between two permeable, parallel discs placed within a magnetic field. Each disc is maintained at a distinct isothermal temperature and is allowed to rotate independently. The upper disc is then set in motion towards the lower one. Using a similarity solution, analytical and numerical solutions are derived from which the effects of various flow parameters on the temperature and heat transfer profiles are studied. The Hartmann number has an effect on the temperature profiles when the discs are allowed to rotate in the same direction, but the effects are negligible during rotation in the opposite direction. The Prandtl number also affects the temperature and heat transfer profiles for small values of the Squeeze Reynolds number. The effect of permeability on the heat transfer profile varies depending on the speed of rotation of the discs relative to each other.
Keywords: heat transfer, permeable discs, electrically conducting fluid.

1 Introduction

The problem of a fluid enclosed between two parallel discs has long been a source of intrigue to many mathematicians over the years. von Kármán was the first to study the situation where an infinite disc rotated in an unbounded fluid [1]. It was Hamza however who examined the effect of a magnetic field on the radial and axial velocity profiles associated with a fluid film squeezed between two rotating surfaces [2]. He later studied the case of irrotational discs subjected to temperature variations and found that the effect of squeezing was of primary significance, whereas the effect of the magnetic forces was negligible [3]. Further, Hamza and

Bhatt [4] and Bhatt and Hamza [5] replaced the solid discs by permeable ones and the radial and axial velocity profiles of viscous fluids squeezed between discs were examined.

In this paper we draw on the ideas of these studies and consider the temperature and heat transfer profiles of a viscous fluid squeezed between two permeable parallel disks rotating in the presence of a magnetic field. We examine the case where the two discs are isothermal with different temperatures.

2 Formulation of the problem

Consider axisymmetric flow of a Newtonian fluid confined between two naturally permeable discs of permeabilities $k_i(1-\alpha t^*), i = 1, 2$ for the lower and upper discs. A magnetic field B(t*) is applied perpendicular to the discs. The discs are rotating in their own planes about the z^* axis of rotation with angular velocities $\Omega_1(1-\alpha t^*)^{-1}$ and $\Omega_2(1-\alpha t^*)^{-1}$ at a distance $h(t^*) = H(1-\alpha t^*)^{\frac{1}{2}}$ at any time t^*. The upper disc approaches the lower fixed disc with velocity $h'(t^*) = \frac{dz}{dt}$. Both discs are maintained at distinct isothermal temperatures.

To satisfy the geometry of the problem we use cylindrical polar coordinates. Using a similarity solution in accordance with Hamza [2] we take the velocity components (u^*, v^*, w^*) and temperature T^* as

$$u^*(y) = \frac{\alpha r^* f'(y)}{2(1-\alpha t^*)}, \quad v^*(y) = \frac{r^* \Omega g(y)}{1-\alpha t^*}, \quad w^*(y) = \frac{-\alpha H f(y)}{\sqrt{1-\alpha t^*}}; \tag{1}$$

$$T^* = \frac{T - T_0}{T_0} = \frac{a(y)}{1-\alpha t^*} \tag{2}$$

$$B = \frac{B_0}{\sqrt{1-\alpha t^*}} \tag{3}$$

where

$$y = \frac{z^*}{H\sqrt{1-\alpha t}}, \qquad r = r^*, \qquad z = z^*, \qquad t = t^* \tag{4}$$

The mass continuity equation is thus satisfied. We now apply the Navier Stokes equations to the system to obtain

$$\begin{aligned} \frac{1}{r}\frac{\partial p}{\partial r} = \frac{\rho \alpha^2}{4(1-\alpha t)^2} \bigg(& \frac{1}{Re^S} f''' - y f'' - 2f' - (f')^2 \\ & + 2f'' f + \left(\frac{Re^R}{Re^S}\right)^2 g^2 - \frac{M^2 f'}{Re^S} \bigg) \equiv P_1(y,t) \end{aligned} \tag{5}$$

$$\frac{g''}{Re^S} - g'y - 2g - 2f'g + 2fg' = 0 \tag{6}$$

$$\frac{\partial p}{\partial y} = \frac{-\rho \alpha^2 H^2}{2(1-\alpha t)} \left(-y f' - f + 2 f f' + \frac{f''}{Re^2} \right) \tag{7}$$

where the Rotational Reynolds number, the Squeeze Reynolds number and the Hartmann number are respectively defined as

$$Re^R = \frac{\rho\Omega H^2}{\mu}, \quad Re^S = \frac{\rho\alpha H^2}{2\mu} \quad and \quad M^2 = \frac{B_0^2\Gamma}{\mu},$$

where μ denotes viscosity, Γ denotes electrical fluid conductivity, ρ denotes density. Next consider the energy equation. Assuming negligible viscous dissipation we obtain

$$\frac{a''}{Re^S Pr} + 2fa' - a'y - 2a = 0, \tag{8}$$

where the Prandtl number, Pr is defined as

$$Pr = \frac{c_p\mu}{K}$$

c_p denotes specific heat at constant pressure and K denotes thermal conductivity. We differentiate and then equate equations (5) and (6) to arrive at

$$f^{iv} = Re^S(3f'' + yf''' - 2ff''') + M^2 f'' - 2\frac{(Re^R)^2}{Re^S}gg' \tag{9}$$

The effect of permeability enters through the slip boundary conditions.

$$\frac{\partial u}{\partial z} = \frac{\sigma_1 u}{\sqrt{k_1^*}}, \quad \frac{\partial v}{\partial z} = \frac{\sigma_1}{\sqrt{k_1^*}}\left(v - \frac{\Omega_1 r}{1-\alpha t}\right), \quad w = 0 \quad \text{on z=0}$$

$$\frac{\partial u}{\partial z} = \frac{-\sigma_2 u}{\sqrt{k_2^*}}, \quad \frac{\partial v}{\partial z} = \frac{\sigma_2}{\sqrt{k_2^*}}\left(v - \frac{\Omega_2 r}{1-\alpha t}\right), \quad w = \frac{dh}{dt} \quad \text{on} \quad z = h(t)$$

where $k_1^* = k_1(1-\alpha t)$ and $k_2^* = k_2(1-\alpha t)$.

We consider isothermal discs maintained at different temperatures.

$$T = T_0 \qquad on \quad z = 0$$

$$T = T_0 + \frac{T_0}{1-\alpha t} \qquad on \quad z = h(t)$$

Thus the boundary conditions are:

$$f''(0) = \lambda_1 f'(0), \quad g'(0) = \lambda_1(g(0) - 1), \quad f(0) = 0, \quad a(0) = 0$$

$$f''(1) = -\lambda_2 f'(1), \quad g'(1) = -\lambda_2(g(1) - s), \quad f(1) = \frac{1}{2}, \quad a(1) = 1$$

where

$$s = \frac{\Omega_2}{\Omega_1} \quad \text{and} \quad \lambda_1 = \frac{\sigma_1 H}{\sqrt{k_1}}, \quad \lambda_2 = \frac{\sigma_2 H}{\sqrt{k_2}} \qquad \sigma_i, i = 1, 2$$

represents slip conditions for the lower and upper discs.

3 Approximate solutions

For small parameter values, $Re^S = o(1)$, $Re^R = O(Re^S)$, $M = O(Re^S)$, a regular perturbation solution is used to obtain approximate analytic results. In this paper, the perturbation solution is modelled after those of Hamza [2, 3], Hamza and Bhatt [4] and Bhatt and Hamza [5].

$$f = f_0 + Re^S f_1 + o(Re^S)^2, \quad g = g_0 + Re^S g_1 + o(Re^S)^2,$$
$$a = a_0 + Re^S a_1 + o(Re^S)^2$$

f_0, f_1, g_0, g_1, a_0 and a_1 satisfy the equations

$$f_0^{iv} = 0$$
$$f_1^{iv} = 3f_0'' + yf_0''' - 2f_0 f_0''' + \frac{M^2 f_0''}{Re^S} - 2\left(\frac{Re^R}{Re^S}\right)^2 g_0 g_0'$$
$$g_0'' = 0$$
$$g_1'' = yg_0' + 2g_0 + 2f_0' g_0 - 2f_0 g_0'$$
$$a_0'' = 0$$
$$a_1'' = Pr(2a_0 + ya_0' - 2f_0 a_0')$$

These equations are subject to the conditions:
$f_0(0) = 0, \quad f_1(0) = 0, \quad f_0(1) = \frac{1}{2}, \quad f_1(1) = 0$
$f_0''(0) = \lambda_1 f_0'(0), \quad f_1''(0) = \lambda_1 f_1'(0)$
$f_0''(1) = -\lambda_2 f_0'(1), \quad f_1''(1) = -\lambda_2 f_1'(1)$
$g_0'(0) = \lambda_1(g_0(0) - 1), \quad g_1'(0) = \lambda_1 g_1(0),$
$g_0'(1) = -\lambda_2(g_0(1) - s), \quad g_1'(1) = -\lambda_2 g_1(1)$
$a_0(0) = 0, \quad a_1(0) = 0,$
$a_0(1) = 1, \quad a_1(1) = 0$
The following solutions arise

$$f_0 = \frac{y}{2A}(-2y^2 B + 3y\lambda_1 C + 6C)$$
$$f_1 = \gamma_1 y^7 + \gamma_2 y^6 + \gamma_3 y^5 + \gamma_4 y^4 + \frac{\gamma_5 y^3}{6} + \frac{\gamma_6 y^2}{2} + \gamma_7 y$$
$$g_0 = \frac{1}{B}(E - Dy)$$
$$g_1 = \frac{\kappa_1 y^5}{20} + \frac{\kappa_2 y^4}{12} + \frac{\kappa_3 y^3}{6} + \frac{\kappa_4 y^2}{2} + \kappa_1 y + \kappa_2$$
$$a_0 = y$$
$$a_1 = Pr\left(\frac{y^5 B}{10A} + \frac{y^4 \lambda_1 C}{4A} + \frac{y^3}{6}\left(3 - \frac{6C}{A}\right)\right.$$
$$\left. + y\left(\frac{-B}{10A} + \frac{\lambda_1 C}{4A} - \frac{1}{6}\left(3 - \frac{6C}{A}\right)\right)\right)$$

where

$$A = 4\lambda_1 + 4\lambda_2 + \lambda_1\lambda_2 + 12, \qquad B = \lambda_1 + \lambda_2 + \lambda_1\lambda_2, \qquad C = 2 + \lambda_2,$$
$$D = \lambda_1\lambda_2(1-s), \qquad E = \lambda_1 + \lambda_2 s + \lambda_1\lambda_2$$

$$\gamma_1 = \frac{-B^2}{70A^2}, \qquad \gamma_2 = \frac{\lambda_1 BC}{20A^2},$$
$$\gamma_3 = -\frac{B}{5A} + \frac{3BC}{10A^2} - \frac{M^2 B}{20Re^S A} - \frac{D^2}{60B^2}\left(\frac{Re^R}{Re^S}\right)^2,$$
$$\gamma_4 = \frac{3\lambda_1 C}{8A} + \frac{M^2\lambda_1 C}{8Re^S A} + \frac{DE}{12B^2}\left(\frac{Re^R}{Re^S}\right)^2,$$
$$\gamma_5 = -6(\gamma_1+\gamma_2+\gamma_3+\gamma_4) - 6(\frac{\lambda_1}{2}+1)\gamma_7, \qquad \gamma_6 = \lambda_1\gamma_7,$$
$$\gamma_7 = \frac{1}{\frac{3\lambda_1}{2}+4}((39+7\lambda_2)\gamma_1 + (27+6\lambda_2)\gamma_2 + (17+5\lambda_2)\gamma_3 + (9+4\lambda_2)\gamma_4),$$

$$\kappa_1 = \frac{4D}{A}, \qquad \kappa_2 = -\frac{6E}{A} - 3\frac{\lambda_1 CD}{AB}, \qquad \kappa_3 = -\frac{3D}{B} + \frac{6\lambda_1 EC}{AB},$$
$$\kappa_4 = \frac{2E}{B} + \frac{6EC}{AB}, \qquad \kappa_5 = \lambda_1\kappa_6,$$
$$\kappa_6 = -\frac{1}{B}(\frac{\kappa_1}{20}(\lambda_2+5) + \frac{\kappa_2}{12}(\lambda_2+4) + \frac{\kappa_3}{6}(\lambda_2+3) + \frac{\kappa_4}{2}(\lambda_2+2))$$

4 Numerical solution

In accordance with Hamza [3], the perturbation solutions derived in Section 3 are valid only when the parameters considered are small. When larger parameter values are considered, the numerical solution of the system of equations must be evaluated.

Matlab was used to solve the two point boundary value problems using a collocation method. The original coupled equations were first expressed as a system of seven first order differential equations. A residual function, which returns zero at the boundary values was also defined. Initial guesses, were evaluated through the analytical solutions derived in Section 3.

The method of solution was first evaluated for reliability using previous studies [3, 4, 5]. In these studies, two-point boundary-value problems were solved using the Numerical Algorithm group (NAG) routine (DO2AGF) which used initial value techniques and Newton iteration.

The numerical solutions of the temperature and heat transfer profiles were investigated when the Pr, M, Re^S and Re^R parameters were altered. This was done by plots of a against y and a' against y respectively. This follows from work done by Hamza [3] where he showed that the Nusselt number which is a function of the a' parameter is indicative of heat transfer. This was also supported by McKibbin and O'Sullivan [6].

Also the effects of changes in permeability and plate rotation on the temperature and heat transfer profiles were observed.

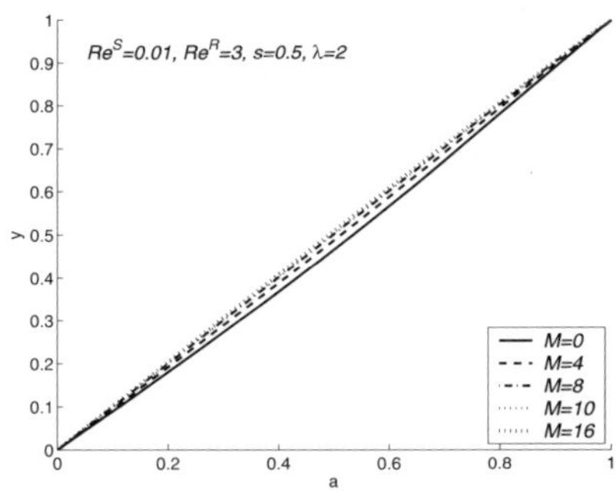

Figure 1: Changes in temperature profile with Hartmann number when discs rotate at different speeds in the same direction.

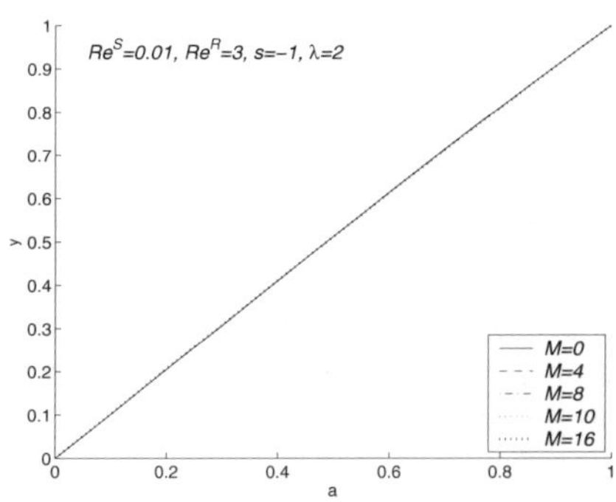

Figure 2: Changes in temperature profile with M when discs rotate at the same speed in opposite directions.

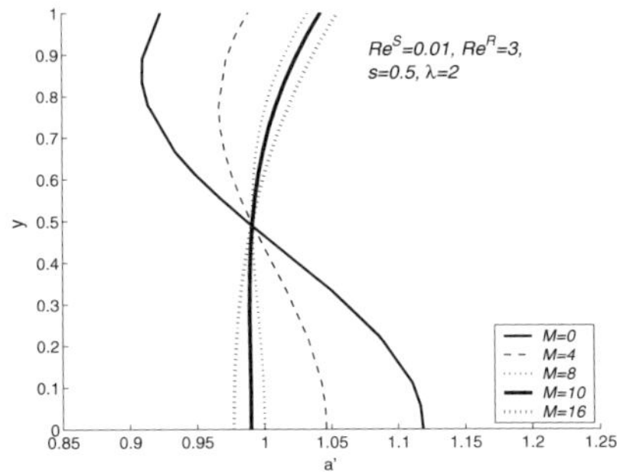

Figure 3: Changes in heat transfer profile with M when discs rotate at different speeds in the same direction.

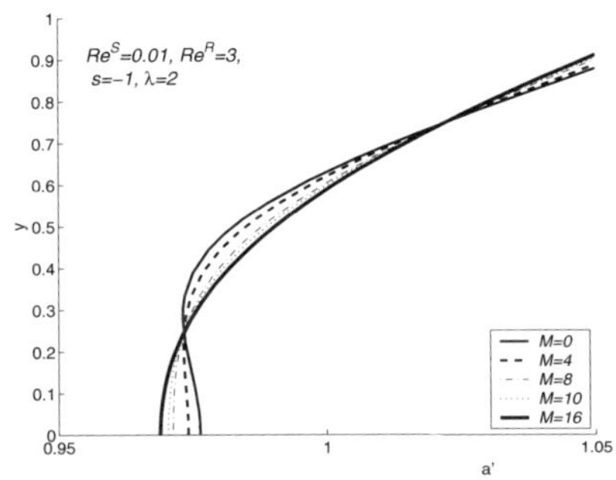

Figure 4: Changes in heat transfer profile with M when discs rotate at the same speed in opposite directions.

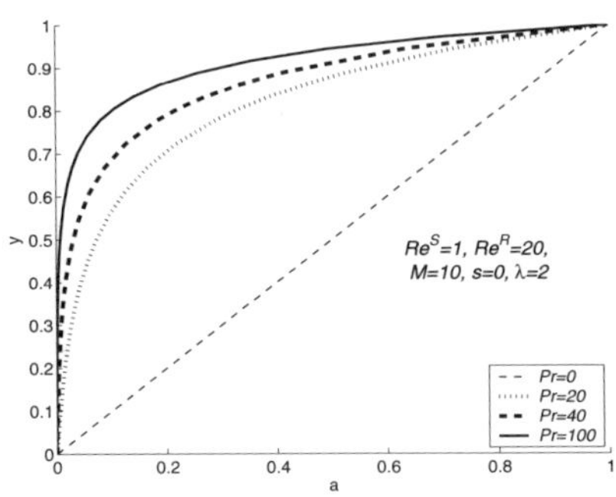

Figure 5: Changes in temperature profile with Pr when discs are irrotational.

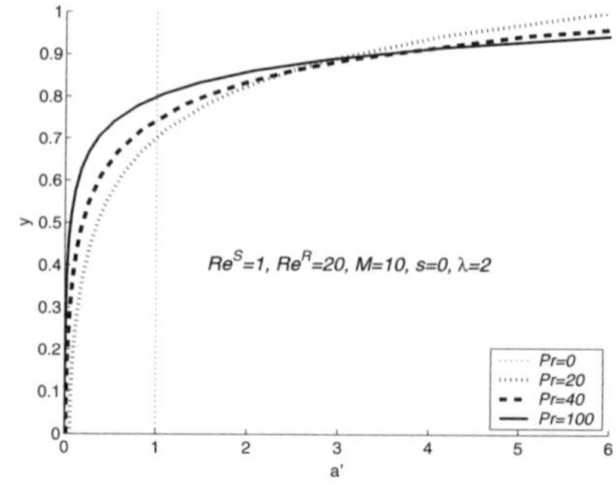

Figure 6: Changes in heat transfer profile with Pr when discs are irrotational.

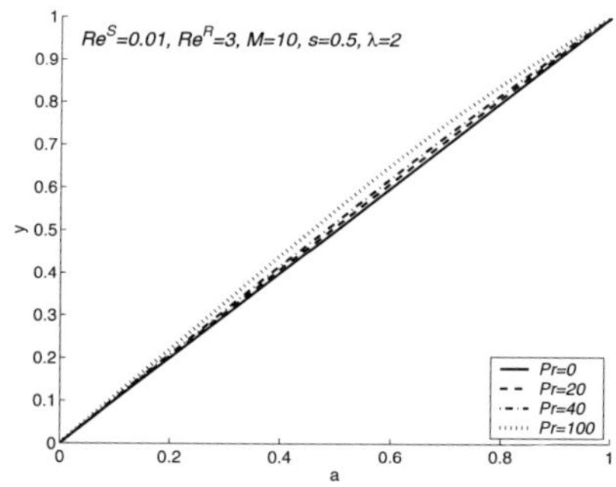

Figure 7: Changes in temperature profile with Pr when discs rotate at different speeds for small Squeeze Reynolds number.

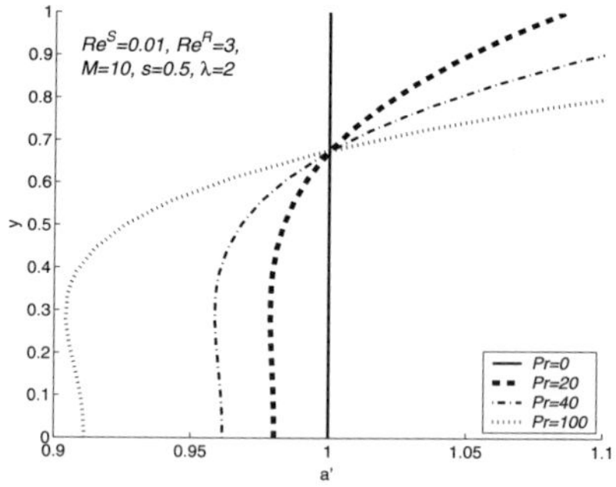

Figure 8: Changes in heat transfer profile with Pr when discs rotate at different speeds for small Squeeze Reynolds number.

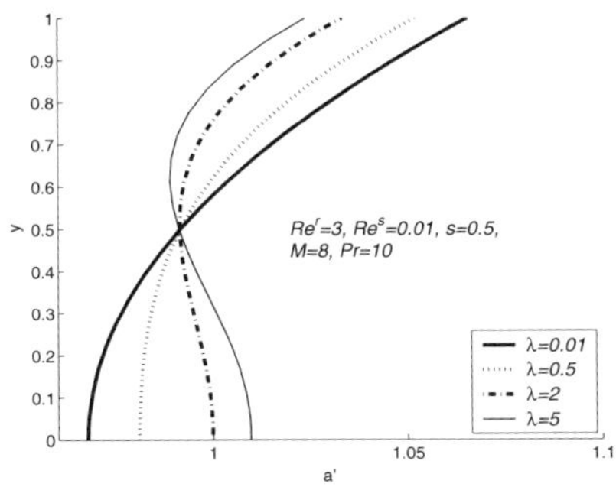

Figure 9: Changes in heat transfer profile with disc permeability with the upper disc rotating at half the speed of the lower one.

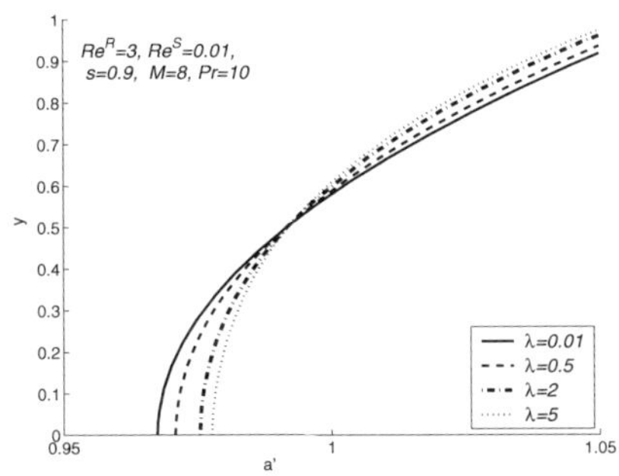

Figure 10: Changes in heat transfer profile with disc permeability for a disc rotation of 0.9.

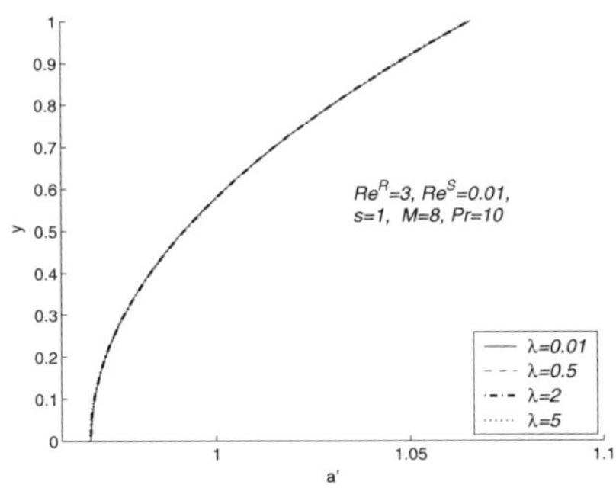

Figure 11: Changes in heat transfer profile with disc permeability for discs rotating at the same speed.

5 Discussion

Figure 1 shows that the temperature profiles are affected by changes in M for the cases $s = 0.5$, Re^S, Re^R and Pr kept constant. This is representative of the discs rotating in the same direction, the upper disc rotating at half the speed of the lower one. A different result occurs when the discs are allowed to rotate in opposite directions with the same speed as is evident in Figure 2. Here it seems that there are negligible variations in the temperature profile with changes in M.

Figure 3 illustrates the effects of the magnetic force on heat transfer. It seems that M affects the heat transfer profiles with strong similarities between the cases where the discs rotate in the same direction, the upper disc rotating at half the speed of the lower one, and where the upper disc is stationary. The profile is different when the discs rotate in opposite directions as shown in Figure 4.

The results are different to those noted by Hamza [3] who studied the case where an upper disc moves towards a lower one, both discs being placed within a magnetic field and maintained at different isothermal temperatures. However in his problem the discs were irrotational.

We also examined the way in which the Prandtl number (Pr) affected the temperature and heat transfer profiles. We considered two main cases. Firstly we used $Re^S = 1$, $Re^R = 20$ and $\lambda_1 = \lambda_2 = \lambda = 2$, fixed $M = 10$ and varied Pr. There were almost identical temperature and heat transfer profiles for the cases where $s = 0$ (Figure 5 and 6) and $s = 0.5$. These profiles showed different characteristics for smaller Reynolds number values, i.e. where $Re^S = 0.01$, $Re^R = 3$ (Figures 7 and 8).

In this problem we have assumed the permeabilities of the upper and lower discs to be small. Recall that large values of λ correspond to small permeabilities, $\lambda_i = \left(\frac{\sigma_i H}{\sqrt{k_i}}\right)$. The effect of λ on the heat transfer profiles becomes significantly less as the rotational parameter s tends to the value 1. This is illustrated in figures 9 through 11.

One area of interest in which this problem can be applied relates to the use of metal lubricants in high temperature porous bearings. They can be used effectively here due to properties like zero-leakage and absence of vibration [7]. There are many other applications of flows driven by rotating discs. Some of these include rotating machinery, lubrication, viscometry, computer storage devices and crystal growth processes [8, 9].

References

[1] von Kármán, T., Uber laminare und turbulente reibung. *Z angew Math Mech*, **1**, pp. 233–252, 1921.

[2] Hamza, E., The magnetohydrodynamic effects of a fluid film squeezed between two rotating surfaces. *J Phys D:Appl Phys*, **24(10)**, pp. 547–554, 1991.

[3] Hamza, E., Unsteady flow between two discs with heat transfer in the presence of a magnetic field. *J Phys D: Appl Phys*, **25**, pp. 1425–1431, 1992.
[4] Hamza, E. & Bhatt, B., Magnetohydrodynamic effects of a fluid film squeezed between two rotating naturally permeable discs: Similarity solutions. *Z angew Math Mech*, **76(10)**, pp. 583–593, 1996.
[5] Bhatt, B. & Hamza, E., Similarity solutions for the squeezed film flow between two rotating naturally permeable discs. *Z angew Math Mech*, **76(5)**, pp. 291–299, 1996.
[6] McKibbin, R. & O'Sullivan, M., Heat transfer in a layered porous medium heated from below. *J Fluid Mech*, **111**, pp. 141–173, 1981.
[7] Shah, R. & Bhat, M., Squeeze film behaviour in rotating porous annular discs lubricated with magnetic fluid. *AJST Science and Engineering Series*, **4(1)**, pp. 32–38, 2003.
[8] Gauthier, G., Gondret, P., Moisy, F. & Rabaud, M., Instabilities in the flow between co- and counter-rotating discs. *J Fluid Mech*, **473**, pp. 1–21, 2002.
[9] Brady, J. & Durlofsky, L., On rotating disc flow. *J Fluid Mech*, **175**, pp. 363–394, 1987.

Section 3
Electrodeposition and formats

Time-dependent simulation of electrochemical machining under non-ideal conditions

D. Curry[1], A. Sherlock[1], A. Mount[2] & R. Muir[2]
[1]*School of Engineering and Electronics, The University of Edinburgh*
[2]*School of Chemistry, The University of Edinburgh*

Abstract

Electrochemical machining (ECM) is a manufacturing process based on the principles of electrolysis. It machines irrespective of workpiece hardness and, as a result, it is used in the manufacture of parts requiring the use of hard alloys such as airfoil blades. Die-sinking ECM involves a shaped tool being moved closer towards the workpiece. Material removal takes place under the influence of an electric field. Once material removal rates are equal to the tool feed rate the process is said to have reached equilibrium. At equilibrium the workpiece is an approximate inverse of the tool shape. Under ideal ECM conditions, a number of process parameters are assumed to be constant. Industrially however, conditions are rarely ideal, and it may not be desirable to machine until equilibrium. This paper presents a time-dependent simulation that includes a model of one of the changing parameters, the overpotential, which is the voltage required to drive the electrochemical reactions at the workpiece and the tool. Previous work has shown that this condition varies throughout the course of machining. Validation is performed against experimental results, and comparisons are made against simulation under ideal conditions.
Keywords: electrochemical machining (ECM), simulation, time-dependent, finite-element analysis, overpotential, non-ideal, non-equilibrium.

1 List of Symbols

V_0	Overpotential	κ	Conductivity
n	Valency	I	Current
V_t	Applied voltage	z	Gap
A	Area of workpiece electrode	$\overrightarrow{dh}$	Material removed

F	Faradays constant = 96500 C mol^{-1}	nJ	Normal current density
		Ω	Subdomain
M	Molecular mass of workpiece	t	Time
		J	Current Density
ρ	Density of workpiece		

2 Introduction

2.1 Background

Electrochemical Machining (ECM) is a manufacturing technique capable of shaping complex parts through anodic dissolution. It is most effectively applied when the material to be shaped is prohibitively hard for efficient machining using conventional contact techniques. It has been adopted in many areas of the manufacturing sector, including the aerospace industry where super-alloys are regularly used. It is a non-contact method that results in little or no tool wear, and ensures that there is no introduction of undesirable sheer stresses in the workpiece material [1, 2]. A variety of products are manufactured using this technique, from turbine blades and blisks in the aerospace industry to microelectronics [3].

This paper discusses die-sinking ECM where the electrolyte is pumped through a gap between the tool (cathode) and the workpiece (anode), removing the products created by dissolution. During the process the two electrodes are fed closer together. After a period of machining, the material removal at the workpiece is equal to the feed rate and an equilibrium-state is reached. At this stage the workpiece is an approximate inverse to the tool shape [1].

2.2 Non-ideal conditions during ECM

Under ideal conditions, a number of process parameters are considered to be constant [1]. Industrially, however, the ECM process often cannot be considered ideal, with variations in process parameters and non-equilibrium machining requirements apparent. In non-equilibrium machining the workpiece is not machined to an equilibrium shape. The tool needs to be designed to produce the correct shape prior to reaching the equilibrium state. An example of such a process is the ECM of the trailing edge of aerofoil blades that are used in turbine engines. Consequently, tool design becomes increasingly complicated, often resulting in the adoption of a costly trial-and-error approach [4]. In an effort to reduce these costs, approaches have been developed to aid the tool design process and simulate machining. Additional work has taken place to include models of varying parameters during machining.

In the area of tool design, analytical approaches have been developed that utilise process parameters to predict a tool-shape from a desired workpiece shape. Methods include simplistic shape projection techniques [1], to more complex conformal mapping techniques [5, 6]. These methods lend themselves to computational efficiency, however they assume equilibrium is reached and can produce unrealistic tool profiles [6].

Time dependent, numerical approaches to simulating the ECM process have also been devised. The finite-element approach has been shown to be a valid way of simulating the process and is capable of producing accurate results [7, 8, 9]. These simulations can offer workpiece shape-solutions during machining, not just at equilibrium, as well as allowing experimental parameters to be monitored over time.

Variations in process parameters can result in the deviation from predicted workpiece shapes calculated assuming ideal conditions. The conductivity (κ) of the electrolyte can vary under the influence of phenomena such as temperature change, flow path variation and hydrogen bubble generation. This variation of κ will alter the rate at which material is removed along the workpiece. Further developments have led to the inclusion of additional mathematical models to simulate some of the process phenomena that are taking place during the ECM process [10]. These include temperature and gas generation within the inter-electrode gap during machining as well as some modelling of the electrolyte flow.

Additionally, variations occur in the chemical parameters that determine the dissolution rate of the workpiece material [11]. These include the valency of the workpiece material and the overpotential required to drive the reactions at each electrode. These parameters have previously been assumed to remain constant during machining.

This paper proposes the inclusion of a parametric model of the changing overpotential into a numerical, time-dependent ECM simulation. It is believed that with the inclusion of this model a more accurate simulation can be produced. The approach demonstrated uses repeated finite-element analysis over a series of time steps allowing geometric solutions at any stage of the machining, not just at the equilibrium state.

3 Overpotential variation during ECM

3.1 Overpotential

During electrolysis two reactions occur: an oxidation process at the workpiece and a reduction process at the tool. The oxidation process results in metal ions being displaced from the workpiece, through an oxide layer, and into the electrolyte solution. The displacement of these metal ions creates vacancies in the oxide layer, which are filled by more ions from the workpiece. The oxide layer is produced by a reaction between the workpiece material and oxygen. At the tool, a reduction process takes place, which releases an hydroxide ion. In ECM, the electrolyte is pumped between the tool and the workpiece in order to remove these ions from the system. In order to drive these electrochemical reactions at tool and workpiece, an overpotential (V_0) is required.

3.2 Characterisation of overpotential

Research has been undertaken to characterise V_0 against measurable machining parameters for a variety of workpiece materials typical to those used industrial

ECM. Typical materials include Inconel, Stainless steel and Titanium. Eqn (1) defines the relationship between V_0 and the current (I) during machining.

$$I = \frac{(V_t - V_0)A\kappa}{z} \tag{1}$$

Ultrasound techniques have been implemented to produce real-time measurement of the inter-electrode gap (z) during machining [12]. This, combined with measurement of I, and knowledge of the tool voltage (V_t), electrode area (A) and electrolyte conductivity (κ), allows the calculation of changes in V_0 during machining.

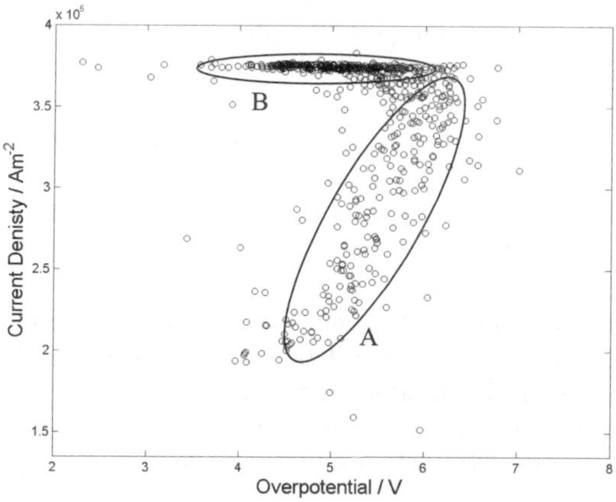

Figure 1: A typical graph showing changes in overpotential against current density. Region A shows the measurements taken before the breakdown of the oxide layer. Region B shows measurements taken once machining through the broken oxide layer.

Whilst machining in region A, shown in Figure 1, the oxide layer is of a constant thickness. The resistance is constant and the layer behaves in an ohmic fashion, producing a linear relationship [13, 14] between V_0 and J. At higher J values, as seen in region B, the rate of dissolution is increased and more ions and electrons are released. The number of vacancies produced increases and they start to coalesce. Consequently, the oxide film begins to breakdown, producing a drop in the resistance, and a reduction in V_0. Experimentally, the tool and the workpiece are moved closer together, reducing the resistance across the inter-electrode gap. Subsequently, J increases and results in an increase of V_0 at the workpiece.

It is the V_0 increase in region A that this paper will introduce as a parametric model into a simulation of the ECM process. Models are still being developed to describe the variations in V_0 seen in region B as well as the transition between the two regions.

4 Simulation overview

The software used for this simulation is MATLAB used in conjunction with the finite-element package FEMLAB. The geometry used represents the electrolyte gap with the upper and lower boundaries defining the profile of the tool and the workpiece. The workpiece boundary is broken down into a number of points, evenly distributed along its length. Figure 2 shows the boundary conditions applied in the model.

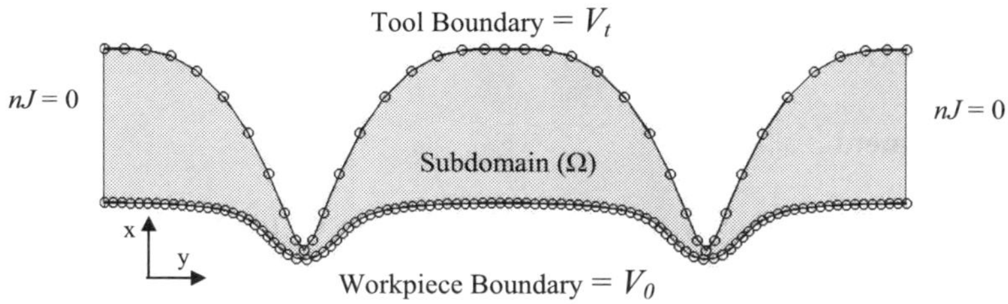

Figure 2: An example of a possible geometry with boundary conditions used in simulation.

The tool boundary is assigned V_t. The side boundaries are constrained so that the normal of the current density (nJ) equals zero. V_0 is applied as the workpiece boundary condition and varies in magnitude in the simulation in accordance with J as discussed in Section 3.2.

The simulation feed rate (f) and V_t are assigned constant values based on typical experimental settings. κ, determined through analysis of experimental electrolyte concentrations, is defined in Ω and the electric field is evaluated through solution of the Laplace field equation,

$$\frac{d^2V}{dx^2}+\frac{d^2V}{dy^2}=0\,. \tag{2}$$

Analysis takes place during each time step (dt), after which the current density (J) data is extracted from points along the workpiece boundary. The material removal at each of these points ($\overrightarrow{dh}$) takes place normal to the workpiece, and is calculated using,

$$\frac{\overrightarrow{dh}}{dt}=\frac{Jtn}{FM\rho}, \tag{3}$$

where n is the valency, M is the molecular mass, ρ is the density of the workpiece, and F is Faraday's constant. A spline representation is used to calculate the normals at each of the points distributed along the boundary.

$\overrightarrow{dh}$ is combined with the tool movement and the points are displaced. A spline is interpolated through the new coordinates, and a new set of evenly distributed points are defined. The geometry is re-formed and re-meshed, V_0 is re-evaluated, the new boundary condition is applied, and solution of the electric field is repeated.

The simulation can be terminated automatically based on either user defined conditions such as elapsed time, or can be dependent upon process parameters being monitored for example current values, or depth of material removed.

5 Validation

5.1 Experimental set-up

A flat-segmented tool was positioned in the base of the machining rig and a 40x15mm Titanium 6/4 workpiece was fed towards it at a rate of 0.5mm/min whilst V_t was set at 16V. Throughout machining a 15% Chloride electrolyte (κ = 20Sm^{-1}) was pumped between the two electrodes at 16 litres/min and recycled through a tank. This flow rate is considered to be sufficiently high to minimise conductivity changes due to electrical heating and products generated during electrolysis. Machining was allowed to take place until equilibrium had been reached. An ultrasound probe is positioned behind the workpiece to measure the workpiece thickness during machining and subsequently allows for calculation of z. Real-time data was collected and stored for later comparison.

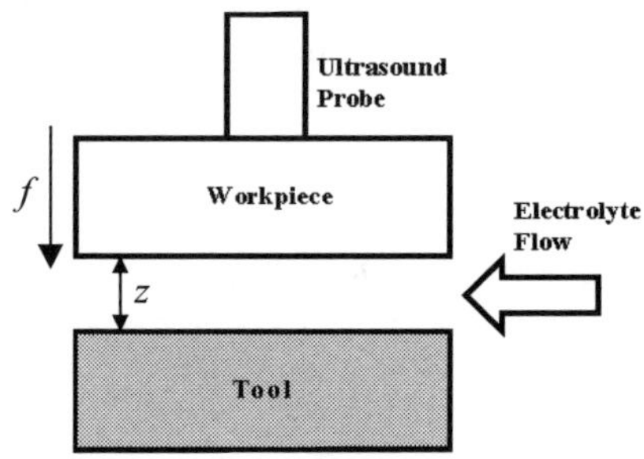

Figure 3: Overview of the machining set-up.

5.2 Simulation set-up

The simulations were performed on a PC with Dual Pentium 4 3.06GHz processor and 2Gb RAM. f and V_t were set to 0.5mm/min and 16V respectively; n was set to experimentally obtained values. Figure 3 shows the V_0 model used for Titanium 6/4 in a 15% Chloride electrolyte. It has been calculated using a least-squares fit through experimental data collected whilst machining in region A.

The simulated machining time was 580 seconds. For further comparison, the same simulation was performed with V_0 = 5.3V, the mean of the measured experimental V_0. This provides a set of results for simulation under ideal conditions.

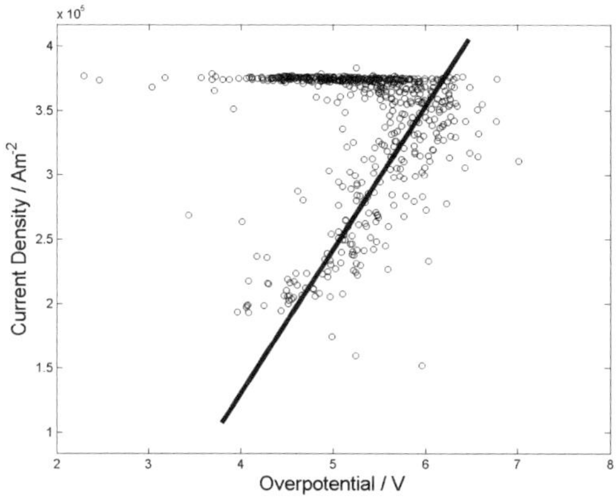

Figure 4: The model (—) used to calculate V_0 during non-equilibrium machining.

5.3 Results

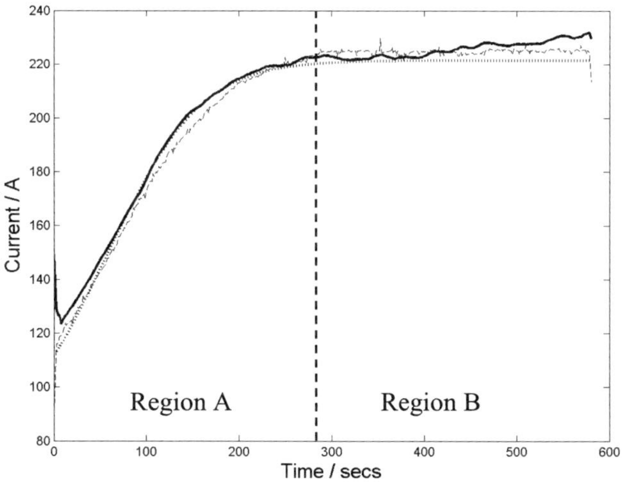

Figure 5: Comparison of I against t for experimental (---), non-ideal simulation (—) and ideal simulation (•••).

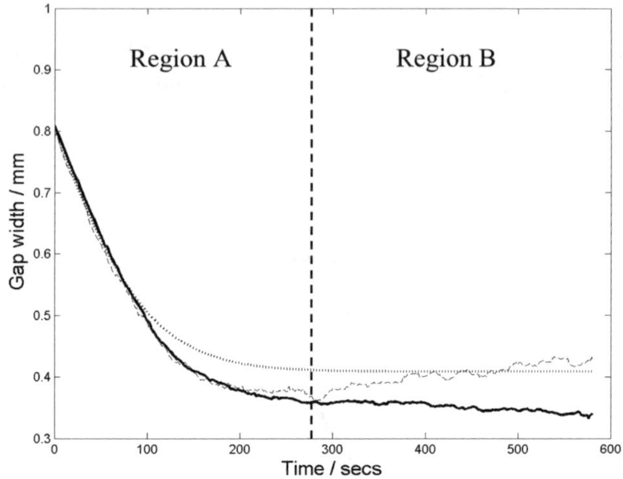

Figure 6: Comparison of z for experimental (---), non-ideal simulation (▬) and ideal simulation (•••).

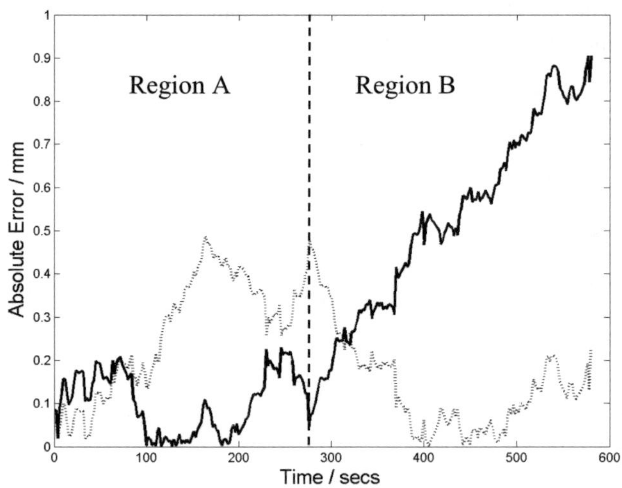

Figure 7: The error between experimental gap and non-ideal gap width (▬) and ideal condition gap width (•••).

6 Discussion

The experimental, non-ideal simulation and ideal simulation results are shown in Figures 5, 6 and 7. Figure 5 and 6 show the current transient and the gap between the tool and the workpiece surface during the machining respectively. Figure 7 shows the absolute error between experimental z and the simulated z for ideal and non-ideal conditions. The figures also show the two V_0 regions, discussed in section 3.2. Region A is defined from experimental data and its duration is the first 270 seconds of machining. After this point, machining is taking place according to region B and the oxide layer is breaking down, resulting in a decrease in V_0.

The current transient, shown in Figure 5, demonstrates that both the ideal and the non-ideal simulations produced similar I values whilst machining in region A. The ideal simulation I becomes constant and the non-ideal simulation I continues to increase. This is due to the continued application of the V_0 model defined for region A throughout the simulation of machining in region B.

The z vs. t graph, shown in Figure 6, shows that the non-ideal simulation closely follows the experimental data whilst machining in region A. Beyond this stage the model used is no longer an accurate representation of the changes in V_0. The ideal simulation z initially matches the experimental data. After 100 seconds however, it indicates that z is wider than that measured experimentally, with deviations of up to 0.5mm, shown in Figure 7.

Once machining in region B, the experimental gap begins to increase as V_0 begins to decrease. Since this region is not modelled in the non-ideal simulation, z continues to decrease. The effect of this can be seen in Figure 7 where the error between experimental and non-ideal simulation z increases. The ideal simulation remains constant throughout region B.

7 Conclusions

A two-dimensional simulation of ECM using finite-element methods has been created that implements a parametric model for overpotential. The simulation is time-dependent, allowing for non-equilibrium machining to be evaluated.

The non-ideal simulation has been demonstrated and the results for simulation of machining using a Titanium 6/4 workpiece has been compared against both experimental data and results from a simulation using ideal conditions. The results for these tests show that the addition of these parameterised values produced a more accurate predication for the current and gap width during non-equilibrium machining states when the oxide layer is of a constant thickness.

Further work is required to develop the overpotential model to include the behaviour of the oxide layer as it breaks down at high current densities, as well as the application of the model to non-flat geometries.

Acknowledgements

The authors would like to thank the British government funding body, EPSRC, and Philips Aerospace for funding this work.

References

[1] DeBarr, A. E. and Oliver, D., *Electrochemical Machining*, 1968 (Macdonald, London).

[2] Wilson, J. F., *Practice and Theory of Electrochemical Machining*, 1982 (Robert E. Krieger Publishing Company, Florida).

[3] Bhattacharyya, B., Mitra, S. and Boro, A. K., Electrochemical Machining: new possibilities for micromachining, *Robotics and Computer Integrated Manufacturing*, **18**, pp. 283-289, 2000.

[4] Jain, V. K., Pandey, P. C., Tooling Design for ECM, *Precis. Eng.*, **2**, pp. 195-206, 1980.

[5] Alder, G. M., Clifton, D. and Mill, F., A direct analytical solution to the tool design problem in electrochemical machining under steady state conditions, *Proc. Instn Mech. Engrs, Part B, Journal of Engineering Manufacture*, **214(8)**, pp. 745-750, 2000.

[6] McClennan, J., Clifton, D., Alder, G. M., Mill, F. and Sherlock, A., Developing two-dimensional electrochemical machining die-sinking tool design and simulation models, *5th International Conference on Behaviour of Materials in Machining*, pp. 307-314, 2003.

[7] Hardisty, H. and Mileham, A. R., Finite element computer investigation of the electrochemical machining process for a parabolically shaped moving tool eroding an arbitrarily shaped workpiece, *Proc. Instn. Mech. Engrs, Part B, Journal of Engineering Manufacture*, **213(8)**, pp. 787-798, 1999.

[8] Hardisty, H., Mileham, A. R. and Shirvani, H., Theoretical and computational investigation of the electrochemical machining process for characteristic cases of a stepped moving tool eroding a plane surface, *Proc. Instn. Mech. Engrs., Part B, Journal of Engineering Manufacture*, **210(2)**, pp. 109-118, 1996.

[9] Jain, V.K., Pandey, P.C., Finite element approach to the two dimensional analysis of electrochemical machining, *Precis. Eng.* **2**, pp. 23-28, 1980.

[10] Kozak, J., Mathematical models for computer simulation of electrochemical machining processes, *J. Mat. Proc. Tech.*, **76**, 170-175, 1998.

[11] Muir, R. N., Mount, A. R., The Dissolution Characteristics of the Stainless Steel Jethete under Electrochemical Machining conditions, *submitted for publication, J Electrochem. Soc.*, 2005.

[12] Clifton, D., Mount, A. R., Alder, G. M., Jardine, D., Ultrasonic measurement of the inter-electrode gap in electrochemical machining, *Int. J. Mach. Tools Manufact.*, **42**, pp. 1259-1267, 2002.

[13] Chao, C. Y., Lin, L. F., Macdonald, D. D., A Point Defect Model for Anodic Passive Films – Film Growth Kinetics, *J. Electrochem. Soc.*, **128**, pp. 1187-1194,1981.

[14] Chao, C. Y., Lin, L. F., Macdonald, D. D., A Point Defect Model for Anodic Passive Films – Chemical Breakdown and Pit Initiation, *J. Electrochem. Soc.*, **128**, pp. 1194-1198, 1981.

Simulation of electrocodeposition

J. Lee & J. B. Talbot
Chemical Engineering Program, University of California, San Diego, U.S.A.

Abstract

The current distributions for various limiting conditions for particle incorporation during electrodeposition of a metal film were numerically investigated. The evolution of a metallic interface is governed by the local current density that is affected by the presence of a non-conducting particle. This study focused on the shape evolution during entrapment of a particle by the growing metal film by mathematical modeling and simulation of the process. The level set method was used to track the moving interface as it grew around the particle, and the governing equation was solved by a finite element method.
Keywords: electrocodeposition, current distribution, simulation.

1 Introduction

Electrocodeposition is the process in which particles are occluded in the plated metallic film; thus, metallic matrix composite films are produced which have been used to enhance wear and corrosion resistance. Electrocodeposition is a function of many process variables, such as bath composition, hydrodynamics, current density, particle characteristics, etc. Stojak et al. [1] have reviewed the general process mechanisms and various models proposed for electrocodeposition. However, the physical phenomenon of electrocodeposition is not well understood. Our experimental research to date has been primarily focused on electrocodeposition under controlled hydrodynamics as a function of process variables, such as current density and bath composition, for copper-alumina composites [2].

The growth of an electrodeposited film along the electrode is controlled by the local current density affected by the presence of an insulating particle during electrocodeposition. The surface and concentration overpotentials are related to the current density normal to the cathode surface. A change in current density affects

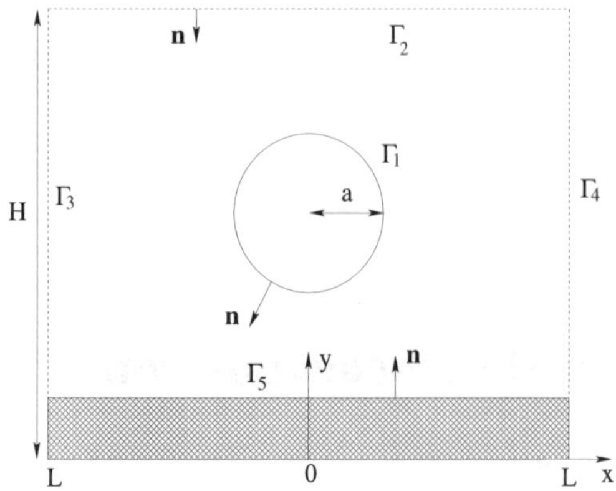

Figure 1: Physical configuration.

the shape of the evolving surface, which may change the microstructure of the electrodeposited film [3, 4]. This study investigates the growth of a metallic film around a single occluded particle as a function of various limiting plating conditions.

2 Formulation of a particle incorporation model

The current distribution during the particle incorporation process is considered in a Nernstian diffusion layer where the effect of lateral migration of the fluid between a particle and an electrode is considered negligible. The metallic interface is initially flat and grows proportional to the normal gradient of current density that is affected by the presence of an insulating particle. The particle radius a is used to make all length scales dimensionless. In fig. 1, L is the characteristic length scale in the x-direction and the normal vectors $\mathbf{n}$ point inward normal to boundaries Γ. The relaxation time of the potential and concentration fields is assumed to be much faster than the time scale of the growth of the electrodeposited film.

The mass balance of the ith species at steady state [5] is given by

$$z_i v_i F \nabla \cdot (c_i \nabla \phi^*) + D_i \nabla^2 c_i = 0, \tag{1}$$

where $v_i (= D_i/RT)$ is the mobility, z_i is the charge number, F is Faraday's constant, c_i is the concentration, D_i is the diffusivity, R is the gas constant, T is the absolute temperature, and ϕ^* is the potential. The first and second terms on the left hand side of eqn. (1) are the contributions of the flux due to migration in an electric field and diffusion due to a concentration gradient, respectively.

When a single deposited species represented by subscript M is considered, eqn. (1) multiplied by $z_M F$ can be recast as the dimensionless concentration c_M normalized by c_∞ and dimensionless electrical potential $\phi(= \phi^* F/RT)$,

$$t_M \Pi \nabla \cdot (c_M \nabla \phi) + \nabla^2 c_M = 0, \tag{2}$$

where $t_M(=z_M^2 v_M c_\infty F^2/\sigma)$ and $\sigma(=F^2\sum z_j^2 v_j c_j)$ are the transfer number of species M and the electrical conductivity of the solution of the bulk electrolyte solution, respectively, and c_∞ is the bulk concentration of species M. The dimensionless parameter Π, is the ratio of the magnitude of the ohmic transfer to that of diffusion transfer,

$$\Pi = \frac{\sigma RT/F}{z_M F D_M c_\infty}. \tag{3}$$

When $\Pi \gg 1$, i.e., the electric field effects are large, primary and secondary current distributions are dominant [4, 6]. When $\Pi \ll 1$, i.e., the effect of mass transfer is much higher than that of ohmic transfer, eqn. (2) is reduced to the Laplace equation for a single species [7].

Periodic boundary conditions are applied on Γ_3 and Γ_4, and on the top boundary Γ_2, that being,

$$c_M = 1 \text{ and } \frac{\partial \phi}{\partial n} = -i_{ave}, \tag{4}$$

where $i_{ave}(=\frac{1}{2L}\int_{-L}^{L} i_n dx)$ is the dimensionless average current density normalized by $\sigma RT/Fa$ and i_n is the dimensionless current density normal to the electrode surface. Since the particle is considered insulating, the fluxes of the concentration of deposited species and the electrical potential on the surface Γ_1 become zero,

$$\frac{\partial c_M}{\partial n} = 0 \text{ and } \frac{\partial \phi}{\partial n} = 0. \tag{5}$$

When the electrical potential on the electrode is assumed to be zero, the dimensionless surface overpotential η_s is given by $-(\phi + \eta_c)$, where the concentration overpotential $\eta_c\left(=\frac{1}{z_M}\log c_M\right)$ is used as a simplification for a supporting electrolyte solution. The nonlinear boundary condition at the interface is derived from the Bulter-Volmer equation,

$$i_n = i_0\left[\exp\left(-\alpha_a\phi\right) c_M^{\beta-\alpha_a/z_M} - \exp\left(\alpha_c\phi\right) c_M^{\beta+\alpha_a/z_M}\right], \tag{6}$$

where α_a and α_c are the anodic and cathodic transfer coefficients and β is the exponent of the composition dependence for the exchange current density [5]. Since the current density at the electrode surface is expressed by the mass flux of species M multiplied by $z_M F$, eqn. (6) can be recast as

$$\begin{aligned} t_M \Pi c_M \frac{\partial \phi}{\partial n} + \frac{\partial c_M}{\partial n} &= \\ -J\Pi\left[\exp\left(-\alpha_a\phi\right) c_M^{\beta-\alpha_a/z_M} - \exp\left(\alpha_c\phi\right) c_M^{\beta+\alpha_a/z_M}\right]&, \end{aligned} \tag{7}$$

where $J(=\frac{i_0}{\sigma RT/Fa})$ is the dimensionless exchange current density.

The dimensionless growth rate of a deposited film V_n normalized by a/τ_D on Γ_5 is obtained by the current density normal to the surface,

$$V_n = -t_M \Pi c_M \frac{\partial \phi}{\partial n} - \frac{\partial c_M}{\partial n} \tag{8}$$

where $\tau_D(= a^2 \rho_M / W D_M c_\infty)$ is the diffusion time scale, W is the molecular weight, and ρ_M is the density of the deposit.

If the time-dependent moving boundary by the zero level set $\psi(\mathbf{x}(t), t) = 0$ is considered, the motion of interface is represented by the chain rule [8]

$$\frac{\partial \psi}{\partial t} + V_n |\nabla \psi| = 0. \tag{9}$$

The moving metallic interface is tracked implicitly by the level set approach which enables the simulation of the entrapment process around the non-conducting particle. The PLTMG (Piecewise Linear Triangle Multi Grid Package) [9] was used to solve eqn. (2). This package is a finite element solver for elliptic partial differential equations in two-dimensions and provided the continuous piecewise linear triangular finite elements and adaptive local mesh refinement. A detailed discussion of the computational procedure for the level set method and PLTMG can be found elsewhere [6].

3 Numerical experiments

The general governing equation and boundary conditions are given by eqn. (2) and (7). Under various limiting conditions, current distributions will be discussed. The values of $\beta = 0.75$, $\alpha_a = 1.5$, $\alpha_c = 0.5$, and $z = 2$ were used in the all computations.

3.1 Primary current distribution

The primary current density distribution depends only on the geometry of the system since the influence of overpotential is negligible. Due to the infinite conductivity of a metallic electrode, the resistance of the electrolyte (ohmic drop) is comparatively much larger than the kinetic resistance due to the surface overpotential. When we consider an insulating particle in contact with a conducting wall, the contact angle between the particle and the electrode causes a change of current density around the contact point; the current density will be infinite when the angle is obtuse and zero when the angle is acute [5].

The insulating particle acts as an obstacle that prevents the current from flowing to the electrode. Additional voltage is needed to sustain the current density around the obstacle and this induces a resistance change, $\Delta V / i_{ave}$, the net voltage change divided by the average current density [6, 10]. The voltage change in the case of a moving film is numerically computed by subtracting the values on the top boundary in the absence of the particle from those in the presence of the particle.

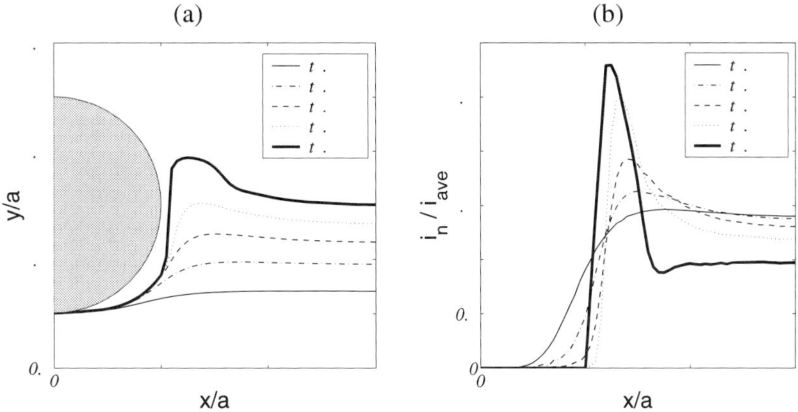

Figure 2: The primary current distribution ($L = 3$): (a) evolution of the metallic interface and (b) normalized current density.

When the particle is incorporated completely by the growing metallic film, the resistance increment will be reduced to zero because of no further voltage change by the presence of the particle.

A flat interface is assumed initially. The normalized current density during the particle entrapment process is presented in fig. 2(b),where $i_{ave} = -0.1$. The maximum value of the normalized current density shifts from the edge of the boundary to near the particle radius, while within the distance $x/a < 1$, the growing velocity reduces to zero in a way that the interface shape deforms largely around the particle, as shown in fig. 2(a). Voids beneath the particle in fig 2(a) are generated by the fast-developed moving metallic film that depends on the characteristics of the potential field [6]. For the primary current distribution, the metallic film does not grow along the non-conducting particle surface. This is the main difference from the other current distributions, which will be discussed. The surface keeps growing after the void is formed, following the gradient of potential fields.

3.2 Secondary current distribution

In general, a more uniformly deposited film can be obtained with a secondary current density distribution [3]. The difference from a primary current density distribution for the model is a nonlinear boundary condition at the electrode surface given by the Butler-Volmer equation which introduces a surface overpotential. Since the dimensionless exchange current density J represents the kinetic rate of the reaction on the electrode, J was used as a parameter in this study to investigate the effect of uniformity and growth of the deposited film [6, 10]. When $\Pi \gg 1$, the concentration gradient is negligible so that c_M is unity, and eqn. (7) is then

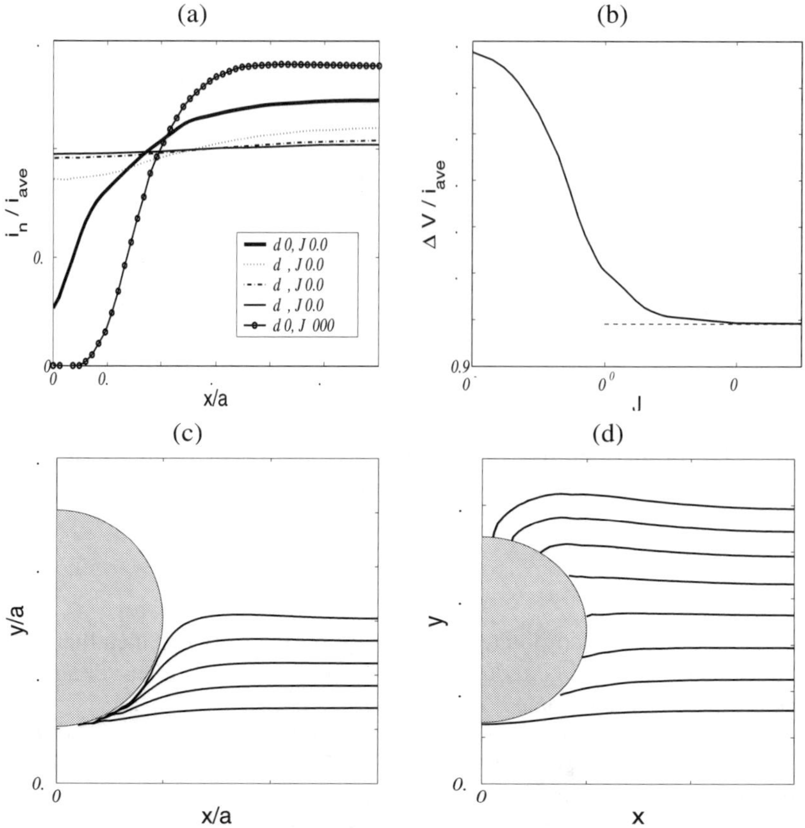

Figure 3: The secondary current distribution ($L = 3$): (a) normalized current density as a function of d and J, (b) resistance change with J, (c) evolution of the metallic interface ($i_{ave} = -0.1$ and $J = 1$), and (d) evolution of the metallic interface ($i_{ave} = -0.5$ and $J = 0.01$).

reduced to

$$\frac{\partial \phi}{\partial n} = -J\left[\exp\left(-\alpha_a \phi\right) - \exp\left(\alpha_c \phi\right)\right]. \tag{10}$$

Figure 3(a) shows the normalized current density along the flat electrode for various values of d, which is the minimum distance between a particle and the electrode, at $J = 0.01$. When a particle comes into contact with the electrode, $d = 0$, the current density even at the contact point is a non-zero value. This makes a more uniformly deposited film compared that for a primary current density distribution. However, as J increases to very large values (towards infinity), the pattern is similar to that of the primary current density distribution. The resistance increment as a function of J for a secondary current density distribution is compared to that for a primary current density distribution at the same condition in fig. 3(b), where the dashed line is for a primary current distribution. As J increases, the resistance increment, $\Delta V/i_{ave}$, decreases and converges to a fixed value of that for a primary current density distribution. However, as the value of J becomes smaller, the kinetic effects of the electrode reaction become important. This is a significant feature as compared with the primary current density distribution. The potential value underneath the particle decreases to compensate for the increased kinetic resistance.

In case of $J = 1$, the growth of the metallic film beneath the particle in fig. 3(c) is slower than that farther away from the particle. This figure shows that the small void between the particle and the deposited film can be explained by both a geometrical effect and the potential characteristics. Figure 3(d) shows the evolution of a growing metallic film for the particle incorporation process. Each line in fig. 3(c) and (d) are regular time steps of 1.5 and 0.5, respectively. The growing metallic film with a small exchange current density is nearly uniform except at the top of the particle. The potential field at the gap between the developing film on both sides of the center line does not change much in the lateral direction; the normal gradient of potential at the metallic interface has nearly vanished. Therefore, the film grows vertically and does not extend into the gap such that a small void is generated on top of the particle. This indicates a possible mechanism for creation of a void even at a small J with a secondary current distribution.

3.3 Tertiary current distribution

For a tertiary current density distribution, the deposited metal film growth is controlled by the rate of surface reaction (kinetic overpotential) and by the mass transfer of reacting species by diffusion (concentration overpotential). When $\Pi \ll 1$, eqn. (2) is reduced into the Laplace equation for a single species and the current density on the electrode surface is only defined by the diffusion flux. Eqn. (7) can be then recast as

$$\frac{\partial c_M}{\partial n} = \xi \, c_M^{\beta+\alpha_c/z_M}, \tag{11}$$

where a single parameter $\xi(= J\exp(\alpha_c\phi))$ has a great influence on the shape change of a growing metallic film [7]. When ξ is large, either the exchange current density is large or the applied potential is large.

Figure 4(a) shows the initial average current density as a function of the polarization factor, where three points (A, B, and C) are chosen for simulation. Each line in fig. 4(b), 4(c), and 4(d) has regular time steps of 5, 2, and 0.5, respectively. As $\xi \to 0$, the surface concentration gradient from eqn. (11) becomes very small and the concentration approaches the bulk concentration. Since the growth rate is very slow, the time for deposition is long, however, a more uniform deposited film is produced, as illustrated in fig. 4(b). At a moderate value of $\xi(= 0.5)$ in fig. 4(c), the deposition time is nearly decreased by four by increasing ξ. When $\xi \to \infty$, the surface concentration goes to zero, and the problem in a mathematical sense is the

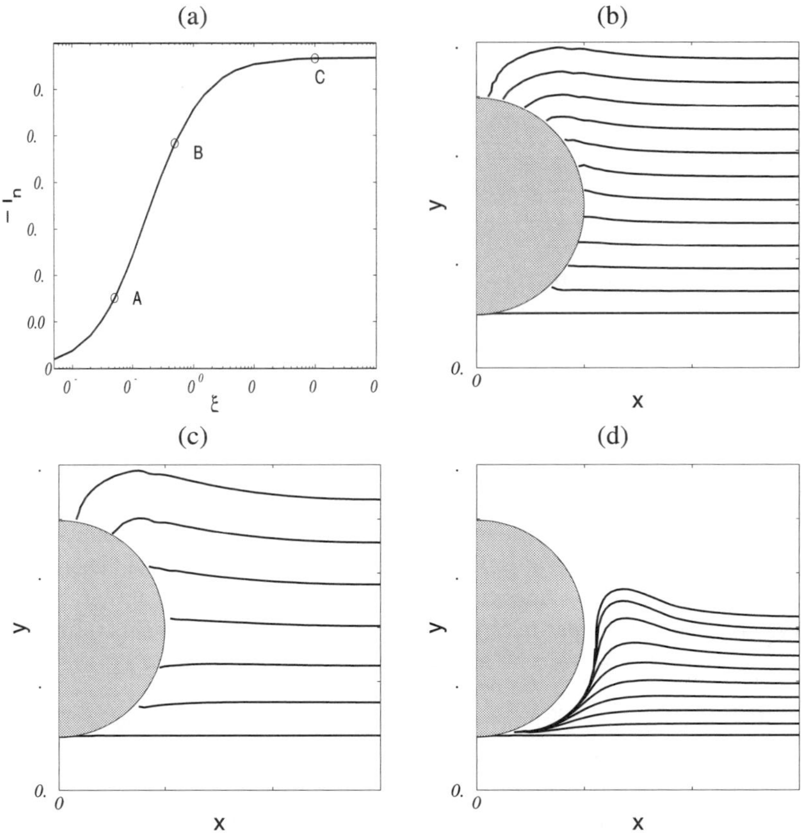

Figure 4: The tertiary current distribution ($L = 3$): (a) average current density vs. ξ, (b) $\xi = 0.05(A)$, (c) $\xi = 0.5(B)$, and (d) $\xi = 100(C)$.

same as the primary current distribution, where ϕ is zero at the metallic surface. A similar evolution shape as in fig. 2(a) is shown in fig. 4(d).

4 Conclusions

The current distributions of a growing metallic interface in the presence of a particle were investigated numerically. For a primary current distribution, the potential characteristics around the particle caused a large deformation of the metallic interface, and finally, created a void below the particle. The kinetic resistance, which is dependent upon exchange current density, with a secondary current distribution resulted in separated potential lines underneath the particle and contributed to a more uniformly deposited film. However, a small void on top of the particle was generated even at a small exchange current density. The tertiary current distribution studied by a single polarization factor allows a more evenly deposited film.

The tertiary current distribution with a conducting particle in the electrocodeposition will be considered in future work. The lateral migration of the fluid will also be considered which will alter the concentration field, which in turn will cause the current density to change.

References

[1] Stojak, J.L., Fransaer, J. & Talbot, J.B., Review of electrocodeposition. *Advances in Electrochemical Science and Engineering*, eds. R.C. Alkire & D.M. Kolb, Wiley-VCH: Weinhein, pp. 193–223, 2002.

[2] Stojak, J.L. & Talbot, J.B., Investigation of electrocodeposition using a rotating cylinder electrode. *Journal of Electrochemical Society*, **146**, pp. 4504–4513, 1999.

[3] Ibl, N., Current distribution. *Comprehensive Treatise of Electrochemistry*, eds. E. Yeager, J.O. Bockris, B.E. Conway & S. Sarangapani, Plenum Press: New York, pp. 239–315, 1983.

[4] Dukovic, J.O., Computation of current distribution in electrodeposition, a review. *IBM Journal of Research and Development*, **34**, pp. 693–705, 1990.

[5] Newman, J.S., (ed.) *Electrochemical Systems*. Prentice Hall: Englewood Cliffs, 1991.

[6] Lee, J. & Talbot, J.B., Simulation of particle incorporation process: Primary and secondary current distributions. *Submitted to Journal of Electrochemical Society*, 2005.

[7] Hume, E.C., Deen, W.M. & Brown, R.A., Mass transfer analysis of electrodeposition through polymeric masks. *Journal of Electrochemical Society*, **131**, pp. 1251–1258, 1984.

[8] Osher, S. & Fedkiw, R., (eds.) *Level Set Methods and Dynamic Implicit Surfaces*. Springer-Verlag: New York, 2003.

[9] Bank, R.E., (ed.) *PLTMG: User's Guide 9.0*. University of California, San Diego, 2004.

[10] Dukovic, J.O. & Tobias, C.W., The influence of attached bubbles on potential drop and current distribution at gas-evolving electrodes. *Journal of Electrochemical Society*, **134**, pp. 331–343, 1987.

Modeling the electrodeposition process: from the macroscopic to the mesoscopic, from continuous mathematics to discrete algorithms

Ph. Mandin[1], J. M. Cense[1], C. Fabian[2], C. Gbado[1] & D. Lincot[1]
[1]*LECA UMR CNRS 7575, ENSCP, France*
[2]*James Cook University, Australia*

Abstract

The chemical bath electrodeposition process is used in many industrial applications to obtain a thin layer material on surfaces. Numerous semi-conductor or magnetic materials, such as oxide, chalcogenure or alloys, are obtained in electrochemical cells on a laboratory scale. Some of these materials are interesting enough to be produced on an industrial scale, for example for fuel cells or photovoltaic applications. In industrial electrochemical cells, dimensions are larger and many properties such as hydrodynamics or electroactive transport are heterogeneous. There are many industrial electrochemical techniques in which the electrode moves with respect to the solution. These systems, like the rotating electrodes, are called hydrodynamic electrochemical processes. It is also interesting to notice that the micronic structures, such as material deposit roughness or porosity, are local properties. The deposit velocity and structure need integrated information from the micronic scale to the industrial scale.

The aim of the present work is to model and to numerically simulate the hydrodynamics and electrochemical coupled phenomenon which occur during the chemical bath electrodeposition process for classical electrochemical configurations. Experimental measurements obtained on a laboratory scale with copper deposits are used to identify a realistic modelling which can be assumed to be conserved during the scale-up. The numerical method used at the continuous scale is the finite volume method. In addition, using a Monte Carlo method, local micronic properties are calculated such as the roughness and the porosity of the thin layer material obtained.

1 Introduction

The present work is interested in the modelling and numerical simulation of the classical electrodeposition process. This process generally implies hydrodynamical and electrochemical properties coupling. Hydrodynamical properties because for laboratory and industrial scales configurations, flow always exist: simply for bulk mixing (magnetic), or to ensure and help the electro-active specie transport at working electrode. At laboratory scale, it is possible using a well-known classical geometrical configuration, like rotating electrode, disk, ring, hemisphere or cylinder. At industrial scale, the configurations are often more difficult: three-dimensional multi-jet, sometimes instationary and turbulent. Electrochemical phenomenon is convective transport dependent. Heterogeneous reactions occur at electrodes and locally change chemical composition from bulk properties. Transport properties like viscosity or specie diffusivity are modified; when the density is also modified, due to heterogeneous reactions, the coupling between mechanical transport and reactive properties can be strong.

Numerous works have been done concerning the electrochemical processes modelling and simulation. Two main approaches are to be cited: the mesoscopic and the continuous. Both need input data and lead to interesting properties. The mesoscopic scale leads to interesting results for electrodeposition processes (Huang and Hibbert [1]) and for crystal growth [2]. The continuous scale, well known in chemical and electrochemical engineering science, have been well investigated by Bard and Faulkner [3], Alden [4], Van den Bossche et al. [5] and many others who have developed nice electrochemical process modelling, simulation and even software tools like Pirode®, monodimensional tool or Myotras®, from Elsyca®.

The aim of the present work is to give some interesting results obtained in case of copper electrodeposition for rotating electrode, disk or cylinder, configurations. Results are mainly obtained at continuous scale, using the Fluent® software and C compilation; large efforts are presently made to couple this approach with mesoscopic scales algorithm to access important micronic structure deposit properties. The present work completes previous work concerning the rotating disk and cylinder electrodes configurations calculations [6], with the copper electrodeposition process results. First, monodimensional results will be used and compared with classical theoretical or experimental results, as validation step. Second, some bidimensional results obtained with the rotating cylinder configuration will be show. A presentation of in-progress works is done, in the last part, of the link between continuous scale results and input data with wanted output properties, like the deposit roughness and porosity and discrete algorithms.

2 From the experiment to the continuous modelling

For the copper electrodeposition process study, the experimentally used electrochemical cell is presented at figure 1. This electrochemical cell is filled with 500 mL of the liquid aqueous electrolyte solution. The rotating cylinder

(Pine Instruments QC3 Series ACQ012CY316) is formed with two parts. One immerged in air and the second which is actually immerged in the aqueous solution with length $z_0 = 5.9\ 10^{-2}$ m. The active cathodic electrode is made with a 316L stainless steel and its centre position can be placed at four positions. At the electrode bottom, to form the classical RDE configuration, and also at three lateral positions which have axial coordinate for centre z_e= 34, 38 or 42 mm. The working electrode has a length $l = 4\ 10^{-3}$ m and a diameter $d_1 = 12\ 10^{-3}$ m (=$2R_1$). The anode is a cylindrical mesh of dimensionally stable anode (DSA) with 13.5 10^{-2} m height, diameter (outer cylinder) $d_2 = 7.6\ 10^{-2}$ m and then the hydraulic diameter is $d_h = 2e = d_2\text{-}d_1 = 64\ 10^{-3}$ m. The reference electrode MSE is the mercurous/mercuric sulphate in saturated K_2SO_4 electrode (651 mV Vs SHE).

For copper electrodeposition process, it is usually used an electrolytic solution, of cupric ions $CuSO_4$ and sulphuric acid H_2SO_4 (supporting salt). The two associated concentration are $C_{Cu}^{tot} = 0.567$ M and $C_{H2SO4}^{tot} = 1.63$ M. Two new organic additives have also been used to produce a better quality copper deposit in terms of purity and smoothness than the standard-industry additive. These additives behave similar to polyethylene glycol (PEG). The experimental study is mainly interested in the deposition rate velocity profile and deposit roughness sensitivity with these additives. The copper deposit surface roughness has been measured and the cross section has been examined using scanning electron microscopy (SEM). These results will not be presented here for confidentiality. In this study it is supposed that copper sulphate salt dissociates completely in water for experiment conditions, according with reaction R1:

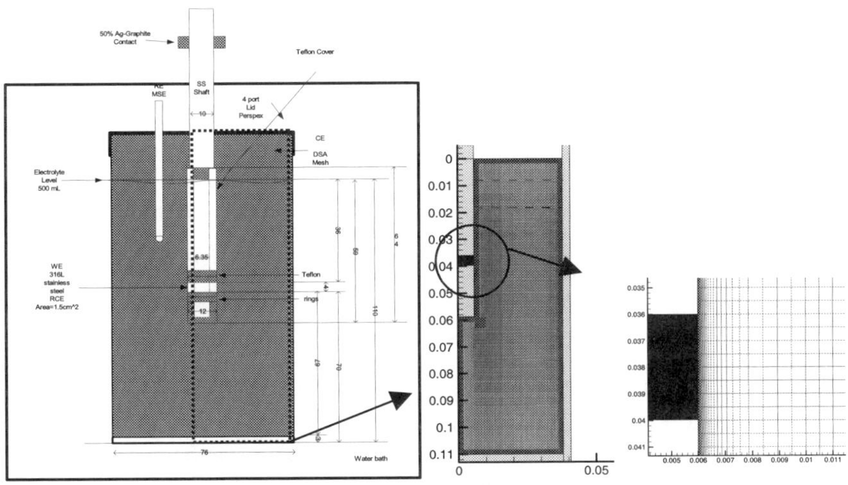

Figure 1: Experimental set-up: geometrical configuration and associated modelling domain with mesh, and zoom of the mesh near electrode.

$$CuSO_4 \Leftrightarrow Cu^{2+} + SO_4^{2-} \qquad \text{(R1)}$$

Concerning heterogeneous reactions, the hydrogen reduction is here neglected, only the electrochemical copper reduction R2 has been considered:

$$Cu^{2+} + 2\ e^- \Leftrightarrow Cu \qquad \text{(R2)}$$

Regarding the works of Selman and Newman [7], Volgin and Davydov [8] or Pirogov and Zelinsky [9], these assumptions are realistic for small pH values, at high-imposed current density.

The investigated factors in the present study have been the temperature T (45 or 64 °C) and the rotation velocity ω (1 or 2.6 rad s^{-1}) have defined four experiments called exp1 to exp4. Experimental and numerical experiments have been done for each combination of factors level choices, according with classical experimental design order 4 Hadamard matrixes [10].

The aqueous solutions considered in the present work have as kinetic viscosity $\nu = 9.2\ 10^{-7}\ m^2\ s^{-1}$ and $6.6\ 10^{-7}\ m^2\ s^{-1}$ and as diffusion coefficient for electroactive specie $D = 9.16\ 10^{-10}\ m^2\ s^{-1}$ and $1.62\ 10^{-9}\ m^2\ s^{-1}$, respectively for temperature $T = 45$ and 64 °C [11]. The Schmidt number ($Sc = \nu / D$) then varies from 1010 to 408.

The viscosity and the diffusion coefficient have been supposed local composition independent, whereas the density sensitivity with local composition is supposed to be independent or with sensitivity $\partial\rho/\partial C = M_{CuSO4}$ the molar weight.

For an incompressible stationary fluid flow in the electrochemical cell must obey both the continuity equation (1), the Navier-Stokes equation (2) and the electroactive specie mass balance (3) with their appropriate boundary conditions:

$$\text{Continuity equation:} \qquad \mathrm{div}(\rho \boldsymbol{V}) = 0 \tag{1}$$

$$\text{Navier-Stokes equation:} \qquad \rho \boldsymbol{V} \mathbf{grad}\, \boldsymbol{V} = \mathbf{div}\, \boldsymbol{\tau} - \mathbf{grad}\, P + \rho \boldsymbol{g} \tag{2}$$

$$\text{Mass balance equation:} \qquad \rho \boldsymbol{V} \mathbf{grad}\, Y = \mathrm{div}\, (\rho D\, \mathbf{grad}\, Y) \tag{3}$$

In these equations, ρ is the density (kg m^{-3}), $\boldsymbol{V}$ is the local velocity flow vector (m s^{-1}), $\boldsymbol{\tau}$ is the shear stress tensor (Pa), P is the local pressure (Pa), $\boldsymbol{g}$ the gravity acceleration vector (m s^{-2}), D the electroactive diffusivity ($m^2\ s^{-1}$) and Y its local mass fraction. Because of the H_2SO_4 supporting electrolyte large concentration, the electromigration transport is here neglected. Under diffusion limited reaction assumption, the electroactive specie concentration at electrode is null, whereas a zero Neuman condition is used for the other frontiers.

Though the hydrodynamic flow is tricomponent $\boldsymbol{V}(z; r) = (v_z(z; r); v_\theta(z; r); v_r(z; r))$, each component, using the axisymetric assumption, is only bidimensional, depending only with $(z; r)$. The flow is said bidimensional axisymetric swirl.

Numerical calculations have been made using the finite volume method and the Simplec algorithm for pressure-velocity coupling of Patankar [12] which is developed in numerical software tool Fluent®. Domain has been meshed using unstructured algorithm, and boundary layers with thinner cell 10^{-5} m, according with calculated phenomena. The power law scheme for discretization has been chosen, with default settings under relaxation coefficients.

3 Monodimensional electrodeposition

The calculations have been validated using the classical Levich theory, which leads to well known properties (see for example Bard and Faulkner [3]):

$v_z = (\omega \nu)^{1/2}(-a\,\gamma^2 + \gamma^3/3 + b\,\gamma^4/6 + \ldots)$ and $v_r = (\omega r)(a\,\gamma - \gamma^2/2 - b\,\gamma^3/3 + \ldots)$

with : $\gamma = z(\omega/\nu)^{1/2}$; a=0.51023 ; b=-0.6159. The calculated profiles obtained with the bidimensional assumption have been compared for each experiment with this validating theory. There is a good agreement between numerical results and theoretical law: lower than 9% for the too small rotation velocity ω = 1 rad s^{-1} (experiment 1 and 2), but better than 3% for experiment 3 and 4 (ω = 2.6 rad s^{-1}). The numerical software can be then considered validated.

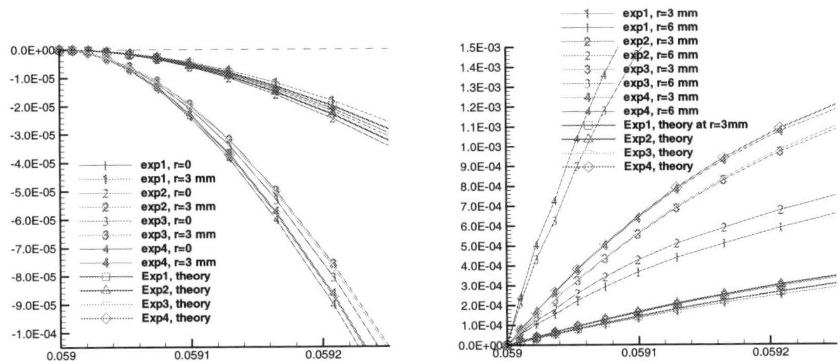

Figure 2: Axial (left) and radial (right) velocity component profiles in the RDE vicinity. Evolution with coordinate z, with temperature (45°C for experiments 1 and 3, 64°C for experiments 2 and 4) and rotation velocity (1 or 2.6 rad s^{-1}).

4 Bidimensional electrodeposition

The second investigated geometrical configuration is the rotating cylinder electrode. Contrary to the previous position, which leads to the 1D RDE configuration, this configuration is actually bidimensional. The use of computational fluid dynamic tool in this case is really adapted. Table 1 gives in first column the experimentally measured limiting current density values j_{exp} (A m^{-2}) for each of the four defined experiments. The aim of this part is the modelling and the prediction of these values. Calculation of the power law sensitivities with explored variables lead to the following experimental correlation: j_{exp} = 8030 $\omega^{0.0471}$ $D^{0.103}$. The sensitivity with rotation velocity is found small whereas, the temperature has, via the diffusivity D, the major influence.

First, literature survey has shown the existence for the RCE configuration of Sherwood correlations, which sometimes use the Reynolds number $Re = 0.5\,\rho\, d_1^2\, \omega / \mu$, like in the Eisenberg work [13], which varies in the present work from about 440 to about 1740; sometimes the Taylor number $Ta = 0.25\,\rho\,(d_2 - d_1)^{1.5}\, d_1^{0.5}\, \omega / \mu$ like in the Mizushina work [14], which varies here varies from about 70 to 280. According with literature results, the hydrodynamic regime is called vortices laminar regime. It exists for these regime two possibilities: without or with periodical transversal oscillations in the circumferential swirl direction.

After a second critic Taylor number evaluated at about 2000, the flow regime is said vortices turbulent. The Mizushina correlation [14], adapted for Taylor number lower than 200 which has been used in table 1, is:

$$Sh = k_j\, d_h/D = 0.74\ Ta^{0.5}\ Sc^{0.33} \quad (4)$$

Table 1: Mass transport properties in term of limiting current density j (A m^{-2}) or Sherwood number ($Sh = k\ d_1/D$) for various modelling assumption.

Exp.	j_{exp}	Re	Ta	Sc	Sh	Sh'	Sh_L	Sh_{num}	j_{num}	j'_{num}	error (%)
1	1130	78	481	1010	163	55	53	54	452	923	-18
2	1544	109	671	408	142	48	46	48	716	1600	3.7
3	1150	203	1250	1010	262	88	83	74	617	926	-19
4	1660	283	1740	408	229	76	71	66	977	1604	-3.3

This mass transport correlation has been obtained using the Chilton-Colburn assumptions, using heat transport measurements at the outer cylinder. Coeuret *et al* [15] (first value in table 1, column *Sh*') and works of Pickett and Stanmore [16] (second value in column *Sh*') leads to propose for the present work electrode configuration the following corrected correlation:

$$Sh' = k_j\, d_1/D = 0.74\ (Ta/2)^{0.5} \times Sc^{0.33} \times (d_h/l)^{0.33} \times (d_1/d_h) \quad (4')$$

Figure 3 shows the calculated stream function ψ for the four experiments. The domain exhibits two contra-rotating vertices, according with Coeuret and Legrand experimental resident time distribution measurements [15].

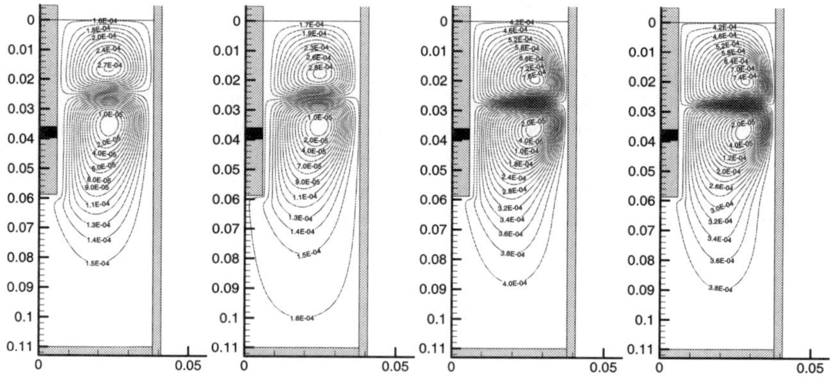

Figure 3: Stream functions (ψ / kg s^{-1}) for ω = 1.0 rad s^{-1} (two left) and ω = 2.6 rad s^{-1} (two right), for reference case at temperature T = 45°C (first and third), and T = 64°C (second and fourth).

The calculation of the flow field and then of the wall shear stress $s = \partial v_z / \partial r = \tau_{zz} / \mu$ at electrode allow the calculation of the mass transport Sherwood number Sh_L using the Lévêque [17] equation (5):

$$Sh_L = 0.807\ d_1\ s^{0.33}\ D^{-0.33}\ l^{-0.33} \qquad (5)$$

Results obtained with this numerical treatment are shown in table 1 and are in good accordance with empirical eqn. (4').

After the calculation of the flow properties, the mass balance for electroactive specie copper Cu^{2+} has been calculated, and leads to the mean Sherwood number and to the mean current density $j = nFDShC_{bulk}/d_1$, with n=2 the electron number and F=96500 C mol^{-1}. It is the non-coupled assumption, which leads to the power law sensitivity $j_{num1} = 5540\ \omega^{0.325}\ D^{0.140}$ which is quite different with the experimental measurement law. The sensitivity with rotation is very strong whereas it is not for experimental results.

The reason is given using the literature results for the copper electrodeposition process for vertical electrodes: for copper electrodeposition, density changes in the electrode vicinity are large, and then, buoyancy effects leads to a natural (also called free) convective transport in the electrode vicinity. Density decreasing from bulk to electrode, according with sensitivity to local concentration already given, is a motion source term.

According with this and the mass conservation eqn. (1), the axial velocity increases at the electrode vicinity, according with figure 4.

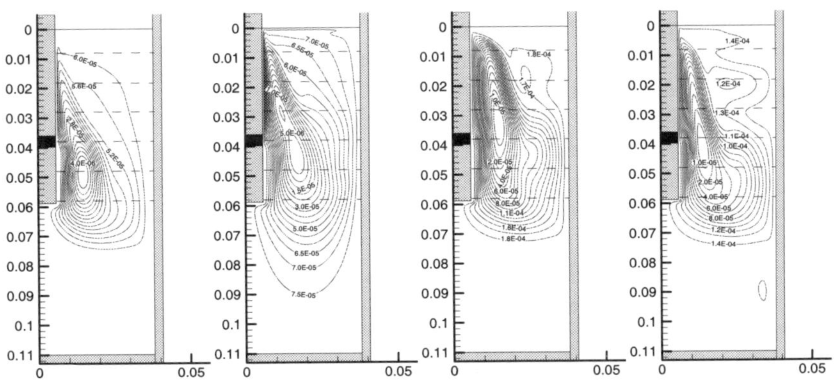

Figure 4: Stream functions with gravity effects (ψ /kg s^{-1}) for ω=1.0 rad s^{-1} (two left) and ω = 2.6 rad s^{-1} (two right), for reference case at temperature T = 45°C (first and third), and T = 64°C (second and fourth).

The electrodeposition process modelling leads this time to a strongly coupling between hydrodynamic properties and chemical composition. The limiting current density have been calculated for each experiment and results given in table 1 leads to a good agreement (better than 4%) between numerical results and experimental measurements for the larger temperature value T=64°C. The complete dissociation assumption for reaction (R1) is realistic for this temperature but not for the smaller value T = 45°C. The simplified chemical-electrochemical modelling has to be improved to obtain better comparison. It is possible for example to use the Pitzer modelling to have a better chemical-electrochemical modelling. Future works for the continuous modelling of the copper elctrodeposition process will take into account such modelling improvement.

In the present work, only limiting density current predictions are shown, but, the continuous scale can also yields to the local deposition rate at electrode for current density smaller values. The present work results have shown the gravity acceleration impact at the continuous scale, mainly upon the coupled flow-composition properties. Fukunaka et al. [18] have also shown a microstructure impact of the gravity field direction and intensity. The aim of the next part is to present the link between continuous scale modelling and discrete algorithm to have microstructural properties prediction.

5 From continuous scale to mesoscopic scale

Prediction of local transport fluxes is generally not sufficient for the electrical use of such thin layers material deposit for industrial applications. The composition, compacity (or porosity) and surface roughness for successive deposition or multilayers contact or for interconnect contact quality are also needed for industrial applications, even if a thermo-mechanical process is used.

It is then decided to complete the continuous scale modelling approach with a mesoscopic modeling. The simplified discrete algorithm used in the present work has been presented by Huang and Hibbert [1]. This work uses a Monte Carlo algorithm. The calculation, according with the previous part, uses the diffusion limited aggregation assumption (DLA) assumption. Matsushita and co-workers have reported that in electrochemical experiment, the cluster is usually fractal and well described by DLA. Dendritic deposit observed during copper electrodeposition is well described by such a modeling which neglects surface diffusion. Same assumptions have been used in the present work to have a qualitative description of such deposit discrete properties.

In this discrete modeling, the conservation equations for reactive species are discretized in the electrode vicinity and relations obtained allow the displacement probability in each of the four directions (2D modelling). In the electrode vicinity, the diffusion coefficient and density are constant, and eq. (3) becomes:

$$D\,\partial^2 Y/\partial z^2 + D\,\partial^2 Y/\partial r^2 - v_z\,\partial Y/\partial z - v_r\,\partial Y/\partial r = 0 \qquad (3')$$

The curvature effects are neglected. The two first terms are associated to diffusive transport and the two last to convective transport. Equations are written in an adimensional form with: $Pe_{z,r} = v_{z,r}\,L/D$, the local Peclet number which is equal to the adimensional velocity; the adimensional lengths are: $Z = z/L$; $R = r/L$ and the discrete algorithm leads to $\delta Z = \delta R = \Delta = L/n_c$, with n_c, the number of cells for each direction, and L the characteristic length.

Then equation (3') can now be written as follows:

$$\partial^2 Y/\partial Z^2 + \partial^2 Y/\partial R^2 - Pe_z\,\partial Y/\partial Z - Pe_r\,\partial Y/\partial R = 0 \qquad (3'')$$

This differential equation is discretized, obeying the finite differences rules, to obtain an algebraic relation between local and neighbouring concentrations:

$$\partial Y/\partial Z = (Y_{i,j} - Y_{i-1,j})/\Delta$$
$$\partial Y/\partial R = (Y_{i,j} - Y_{i,j-1})/\Delta$$
$$\partial^2 Y/\partial Z^2 = (Y_{i+1,j} + Y_{i-1,j} - 2\, Y_{i,j})/\Delta^2$$
$$\partial^2 Y/\partial R^2 = (Y_{i,\, j+1} + Y_{i,j-1} - 2\, Y_{i,j})/\Delta^2$$

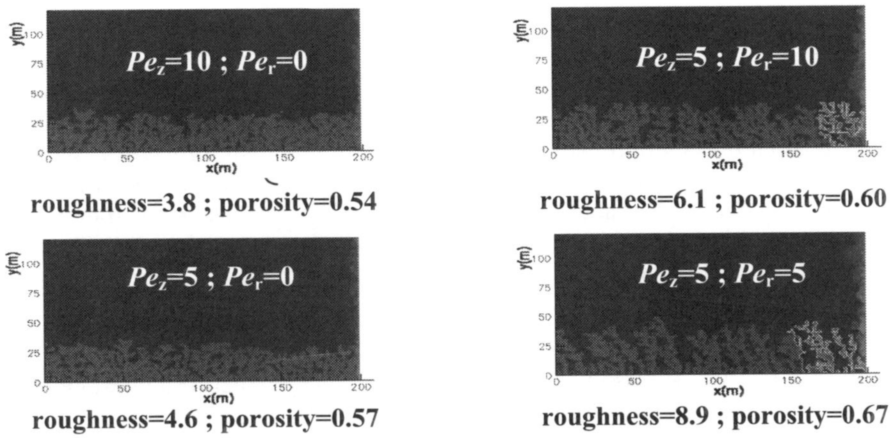

Figure 5: Microstructure calculation for the four experiments at axis (r=0 mm, at left) and at r=3 mm (right) for the RDE configuration.

The i and j integer indexes are associated and orientated with the local axial and radial flow velocity components to ensure positive probabilities. Substituting these differentiations in equation (3''), it can be written:

$$Y_{i,j} = [Y_{i+1,j} + (1+\Delta Pe_z)\, Y_{i-1,j} + Y_{i,j+1} + (1+\Delta Pe_r)\, Y_{i,j-1}]/[4 + \Delta(Pe_z + Pe_r)]$$

The algebraic relation between local concentration $Y_{i,j}$, neighboring concentrations Y, and the displacement probabilities for the 2 directions and 4 senses, are:

$$Y_{i,j} = p_{i+1,j}\, Y_{i+1,j} + p_{i-1,j}\, Y_{i-1,j} + p_{i,j+1}\, Y_{i,j+1} + p_{i,j-1}\, Y_{,j-1} \qquad (3''')$$

The Monte Carlo particle displacement probability is related to the local ratio between convection and diffusion in z and r direction obeying:

In the axial flow direction: $p_{i-1,j} = (1+\Delta Pe_z)/(4+\Delta Pe_z+\Delta Pe_r) = p(i+) = p(v_z)$

In the radial flow direction: $p_{i,j-1} = (1+\Delta Pe_z)/(4+\Delta Pe_z+\Delta Pe_r) = p(j+) = p(v_r)$

For the two (axial and radial) anti-flow directions:

$$p_{i+1,j} = p_{i,j+1} = 1/(4+\Delta Pe_z+\Delta Pe_r) = p(i-) = p(j-)$$

For pure diffusion, displacement probability is isotropic. The microstructure calculation uses a random walker algorithm, with two parameters input for probabilities. Results shown in figure 5 have been obtained with 3000 simulated particles in a 200^2 grid. Particles are injected at the top of the box, with a random initial position. The way is randomly calculated till the particle is fixed, touching another particle. A cyclic boundary condition has been used for boundaries.

Roughness and porosity values for each case have been reported below each figure. This modeling has to be improved, because a constant probability is used for particles in the whole box. In fact, at the boundary layer frontier, the Peclet number is large and the convective transport is dominant. At the working electrode, the Peclet number is zero: the diffusive transport is dominant. Then, probabilities have to be modified, depending with the particle position. Particle surface transport also can occur which has not been considered here.

6 Conclusion

The modeling, prediction and optimization of the electrodeposition process, in term of homogeneity and microstructure has to use two modeling families which use as input different properties and both leads to important properties. What is difficult but important, it is the validation of a multi-scale modeling, which has to relate random walker algorithms input parameters with calculated properties at continuous scale. Often, a simple post-processing can be employed, though the coupling can be sometimes strong between scales.

References

[1] Huang W., Hibbert D.H., Computer modelling of electrochemical growth, *Phys. Rev. E*, 3rd series, 53, 1 part B, 1996.

[2] Levi A.C., Kotrla M., Theory and simulation of crystal growth, *J. Phys. Condens. Matter*, 9, pp299-344, 1997.

[3] Bard A., Faulkner L.R., *Electrochemical Methods: Fundamentals and Applications*. 2nd edition. John Wiley & Sons, 1991.

[4] Alden J.A., Phd thesis, Oxford University, U.K., 2002.

[5] Van den Bossche B., Bortels L., Deconinck J., Vandeputte S., Hubin A., Journal of Electroanal. Chem., 397, 35, 1995.

[6] Mandin Ph., Pauporté Th, Fanouillère Ph., Lincot D., Journal of Electroanalytical Chemistry 565, 159, 2004.

[7] Selman J.R. and Newman J., *J. Electrochem. Soc.*, 118, 1070, 1971.

[8] Volgin V.M. and Davydov A.D., *Electrochimica Acta*, 49, 365, 2004.

[9] Pirogov B. Ya., Zelinsky A.G., *Electrochimica Acta* 49, 3283, 2004.

[10] Goupy J., *Introduction aux plans d'expériences*, 2nd edition, Dunod, 2001.

[11] Uceda, D., Determination of Mass transfer Characteristics in the Electrolysis of Copper, PhD Thesis, University of Missouri-Rolla, 73, 1988.

[12] Patankar S., *Num. Heat Transfer and Fluid Flow*, Taylor & Francis, 1980.

[13] Eisenberg M., Tobias C.W., Wilke C.R., *J. Electroc. Soc.*, 101, 306, 1954.

[14] Mizushina T., *The Electrochemical Method in Transport Phenomena, Advances in Heat Transfer*, Academic press, 7, 87, 1971.

[15] Coeuret F., Legrand J., *J. Appl. Electrochem.*, 10, 785, 1980.

[16] Pickett D.J., Ong K.L., *Electrochimica Acta* 19, 875, 1974.

[17] Lévêque M.A., Ann. Mines, serial 12, 13, 201, 1928.

[18] Fukunaka Y., Dendritic growth of copper under microgravitational field, *Proc.of the 203rd E.C.S. Congress*, San Antonio, May 2004.

Heat transfer analysis of the EDM process on silicon carbide

R. Mahdavinejad
Mechanical Engineering Deptartment, University of Tehran, Iran

Abstract

The Electro discharge machining (EDM) process is based on the generation of a plasma channel, its growth and the creation of a boiling crater on the workpiece surface. Material removal is mainly due to the presence of thermodynamic forces. Therefore, the process is under the influence of the heat transfer behaviour of the plasma channel and the workpiece. In the heat transfer mechanism, heat conduction is known to be the main major mode. Besides, there is another heat source within the silicon carbide body due to its electrical resistance and it results in Joule Heating. The physical properties of SiC material such as its electrical resistance and thermal conductivity are highly affected by temperature. This paper deals with the solution of the heat conduction and voltage drop equations within the workpiece in a cylindrical coordinate system on a finite cylinder regarding the circular plasma heat source. The finite difference method is used to solve these equations. Calculations show that voltage drops within the SiC body are higher than ones expended on the workpiece surface by a plasma channel. Results also show that an increase in discharge current increases the voltage drops and heat generation within the workpiece body while affecting deeper areas.
Keywords: heat transfer, electro-discharge machining, silicon carbide, plasma channel, finite difference.

1 Introduction

Silicon Carbide (SiC) is the fourth hardest material in the world only after diamond, Boron nitride and Boron carbide. It is widely used in industry due to its special specifications as described in Table 1.

Its high hardness, low conduction of electricity and heat, and also good heat shock resistance which make it suitable for use in furnace as heat-element, in

WIT Transactions on Engineering Sciences, Vol 48, © 2005 WIT Press
www.witpress.com, ISSN 1743-3533 (on-line)

atomic centres as protector and regulator of neutrons, and in machining as an abrasive material [1]. Acheson produced SiC for the first time in 1982 by an industrial method [2]. REFEL SiC that is a special kind of Silicon Carbide has a better heat conduction due to its special production process that makes it of more widely usage. The machining of SiC by conventional methods, with regards to its high hardness, is not possible [3]. Among non-traditional machining methods, EDM is the best one of SiC machining [4].

Table 1: Some characteristic of steel and ceramic materials.

Material	Density (g/cm^3)	Hardness (H.V)	Thermal Expansion 1.0E-6/°c	Thermal con. (W/m.°c)		Spec. Heat (J/g.°c)	Electrical Res. (Ω.cm)
				100°	1200°		
REFEL SiC	3.10	2500	4.30	83.6	38.9	670-710	0.42 (At 25°c) 0.016 (At 1200°)
Hot Pressed Silicon Nitride	3.20	2500-3500	3.20	17.5	14	-	-
Hot Pressed Alumina	3.90	2500	9.0	8.4	5	-	-
Reaction bonded Silicon Nitride	2.60	900-1000	3.2	15	14.2	-	-
Tungsten Carbide (6% Co)	15.0	1500	4.9	86	-	205	20
Steel	7.86	800	12.2 (At 100°c) 13.4 (At 1000°)	51.9	29.7	418	1.0E-5

ED machining method is widely used for making tools, dies and other complex-shaped parts [5]. This process is based on the erosive effects of discharges between the electrodes that are immersed in a dielectric tank. The basic principles of the fundamental theories have suggested that the mechanism is based on a thermal conduction phenomenon governed by heat generated from arc channel and dissipated into the tool and workpiece [6]. As shown in Figure 1, tool and workpiece electrodes, are closed to each other and the space between them is filled with dielectric fluid. Between the nearest opposite peaks of electrodes, non-continues discharges take place and a small part of workpiece surface is removed in each pulse. Finally the supplementary face of tool is made on the workpiece.

In this non-traditional machining method, the behaviour of this semi-conductor material is different from conductors. Thermodynamically mechanism is the main mode in this process that is aligned with creation of melting spots and bulk boiling phenomenon. The high temperature gradients generated at the gap during electrical discharge machining result in large localized thermal stresses in a small heat-affected zone [7]. These thermal stresses can lead to micro-cracks, decrease in strength and fatigue life and possibly catastrophic failure [7]. Although an enormous amount of research effort has been put into representing the EDM process by experimental methods, a more elaborate semi-empirical model, based on thermal-mechanical and statistical approaches, has not as yet been reported [8]. The present paper attempts to overcome this shortcoming.

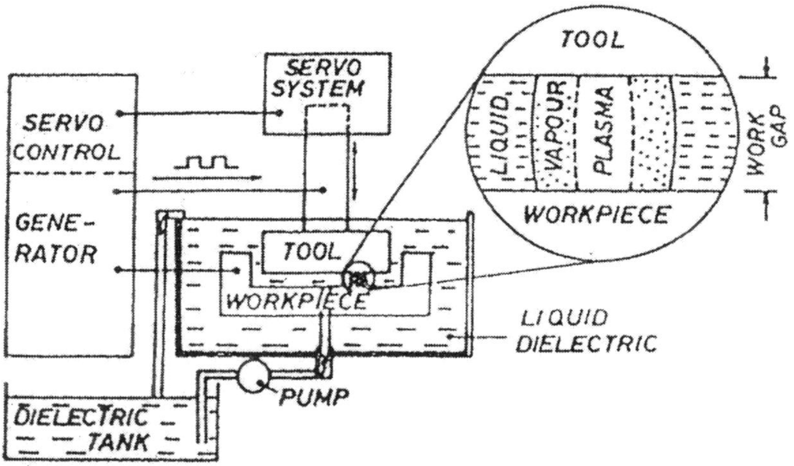

Figure 1: Electro discharge schematic machining process.

2 Principles of heat conduction in EDM

Heat transfer in EDM is mostly a part of plasma-channel heat conduction that reaches to the workpiece and almost all of this is distributed in the workpiece body by heat conduction process [9]. The other heat transfer mechanisms are neglected, since their effects are very small in comparison with heat conduction into the workpiece [10]. The common differential equation of heat conduction in the workpiece is as follow:

$$\nabla .(k\nabla T) + q\circ = \rho c \frac{\partial T}{\partial r} \tag{1}$$

Where k represents heat-conduction coefficient (W/m °C), q° is the rate of internal heat generation (W/m^3), ρ refers to density (Kg/m^3), and c stands for specific heat (J/kg °C). So far various models have been used to solve this conduction equation that almost all of them are for conductive materials in which, the internal Joule heat term, are very small and neglected. To solve the heat-conduction equation in EDM about six models [9] have been suggested that the point source with time dependant radius on a semi-infinite cylinder model is used in this paper.

3 Equations in point heat source model with time dependent radius on a finite cylinder in cylindrical coordinate

In this model, the area of heat source is increased with pulse durations and as the heating flux is constant, its density is decreased (See Figure 2). The differential equation of heat conduction in this model is as follows:

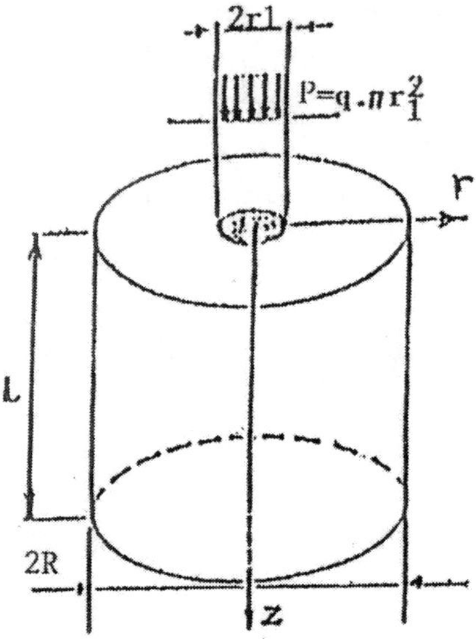

Figure 2: Point sourse model with time dependent radius on a limit cylinder.

$$\frac{1}{r}\frac{\partial}{\partial r}(k.r\frac{\partial T}{\partial r})+\frac{\partial}{\partial z}(k\frac{\partial T}{\partial z})+q\circ(r,z,t)=\rho.c\frac{\partial T}{\partial t} \tag{2}$$

The boundary conditions of this equation are:

$$T(R, z, t) = T_o \tag{3}$$

$$T(r, L, t) = T_o \tag{4}$$

$$K\frac{\partial T(r,o,t)}{\partial z}=0 \qquad r > r_1 (t) \tag{5}$$

$$K\frac{\partial T(r,o,t)}{\partial z}=q \qquad r \leq r_1 (t) \tag{6}$$

And initial conditions are:

$$T(r, z, o)= T_o \tag{7}$$

Where R is the cylinder's radius of solution field (m), L is the length of cylinder on solution field (m), T_o is the ambient temperature ($^\circ$C), $r_1\,(t)$ represents the heat source radius (m), and q stands for the heat flow density (W/m^2).In this model it is assumed that the melting and evaporation processes do not take place. In eqn(2), q° is the rate of Joule heating per unit volume, and its value depends upon the pulse current and the electrical coefficient resistance of the workpiece at any point. The above equation is solved by numerical method in a body graded to small elements. In numerical solution of this equation, although the specifications in each element are assumed to be constant, but the variation of the specifications in different elements are considered.

4 Numerical solution method

To solve the differential equations of heat transfer and electrical potential, the numerical method of finite difference with central difference and explicit method is used. For this purpose, the cylindrical coordinate model has been meshed in radial, vertical and tangential directions. As these two differential equations in environmental direction are symmetric, the equations are only solved for one of the planes passing the vertical axis and cylindrical radius. The meshing is of monotonous kind and the distances between grids are equal. The form of finite difference of electrical potential is as follows:

$$\frac{1}{(\Delta r)^2}[(1-\frac{1}{2i})U_{i,k}+(1+\frac{1}{2i})U_{i+1,k}]+\frac{1}{(\Delta z)^2}(U_{i,k-1}-2U_{i,k}+U_{i,k+1})=0 \tag{8}$$

Also the form of finite difference of Joule heating term is as follows:

$$J_R=\frac{[U_{i-1,k}-U_{i+1,k}]}{2\rho_e.\Delta r}\ ,\qquad J_z=\frac{[U_{i,k-1}-U_{i,k+1}]}{2\rho_c.\Delta z} \tag{9}$$

where U, J_R and J_Z are electrical potential (V) and current densities (A/m^2) in r and Z directions respectively.

By calculation of Joule heating in bulk unit and substituting it in differential equations, we can obtain the explicit form of heat conduction differential equation. In this equation, which is solved for an element, the physical properties of the material are to be constant, therefore:

$$T_{i,k}^{n+1}=\frac{\alpha.\Delta t}{(\Delta r)^2}[(1-\frac{1}{2i})T_{i-1,k}^n+(1+\frac{1}{2i})T_{i+1,k}^n]+\frac{\alpha.\Delta t}{(\Delta z)^2}[T_{i,k-1}^n+T_{i,k+1}^n]+ \\ [1-\frac{2\alpha.\Delta t}{(\Delta r)^2}-\frac{2\alpha.\Delta t}{(\Delta z)^2}]T_{i,k}^n+\frac{\rho_e J^2\alpha.\Delta t}{k} \tag{10}$$

$$T_{0,k}^{n+1} = \frac{4\alpha.\Delta t}{(\Delta r)^2} T_{1,k}^{n} + \frac{\alpha.\Delta t}{(\Delta z)^2}(T_{0,k-1}^{n} + T_{0,k+1}^{n}) + (1 - \frac{4\alpha.\Delta t}{(\Delta r)^2} - \frac{2\alpha.\Delta t}{(\Delta z)^2})T_{0,k}^{n} + \frac{\rho_e J^2 \alpha.\Delta t}{k} \tag{11}$$

$$\alpha = \frac{k}{\rho.c} \tag{12}$$

where α is thermal diffusivity (m^2/s) and n is time index.

Table 2: Some input and output machining parameters.

No. Of Tr.	Material	O.p.c.voltage (v)		$V_{dis}+V_{sic}$ (v)	Req(Ω)		V_{dis} (v)	V_{sic} (v)	I_{dis} (A)	η
		High	Low		Low	High				
4	Si C	216	70	57	2.69	41.67	31.4	25.6	7.90	163
	Steel	216	70	32	2.69	41.67	32	-	18.5	-
6	Si C	216	70	60	1.52	41.67	31.8	28.2	9	177
	Steel	216	70	32	1.52	41.67	32	-	29.4	-
8	Si C	216	70	62	1.06	41.67	33-5	28.5	9.36	170
	Steel	216	70	32	1.06	41.67	32	-	40.3	-

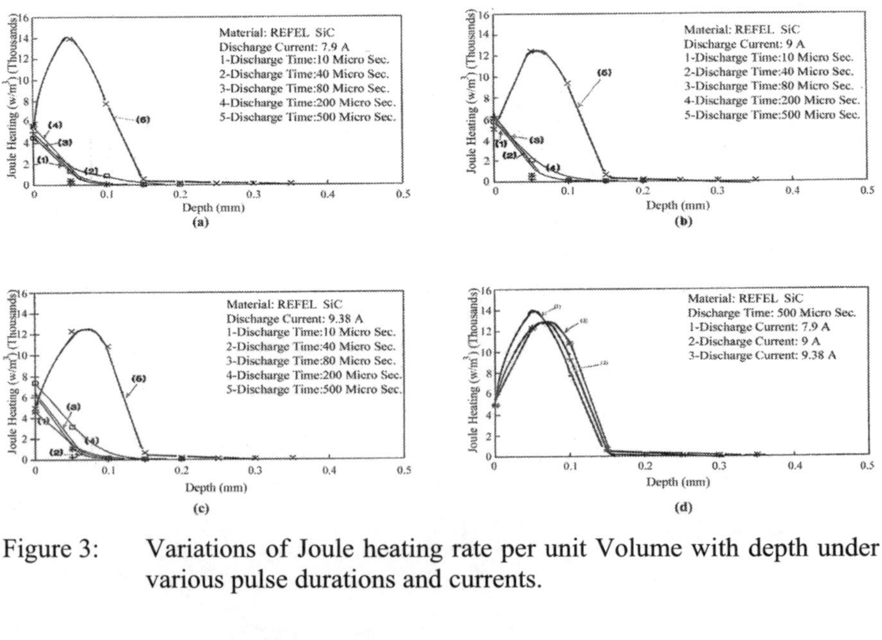

Figure 3: Variations of Joule heating rate per unit Volume with depth under various pulse durations and currents.

5 Discussion

The loss of energy in workpiece as Joule heating is more than 160 percent of plasma energy on the workpiece (See Figure3 and Table 2). This amount is produced mostly near the plasma source and the more discharge current and pulse duration, the more the Joule heating production. The curves of Figure3 also show that in short pulse durations, the maximum rate of Joule heating produced on the workpiece surface, but when the pulse duration increases, this maximum amount is transferred below the surface layer.

It can be seen from Figure 4 that the most temperature changes are on the surface of the workpiece and at a distance of less than 0.2 mm in steel and 0.4 mm in SiC from the centre of melting zone. The curves of Figure 4 also show that the longer the pulse duration, the higher the maximum temperature changes and the more the radius of high temperature zone.

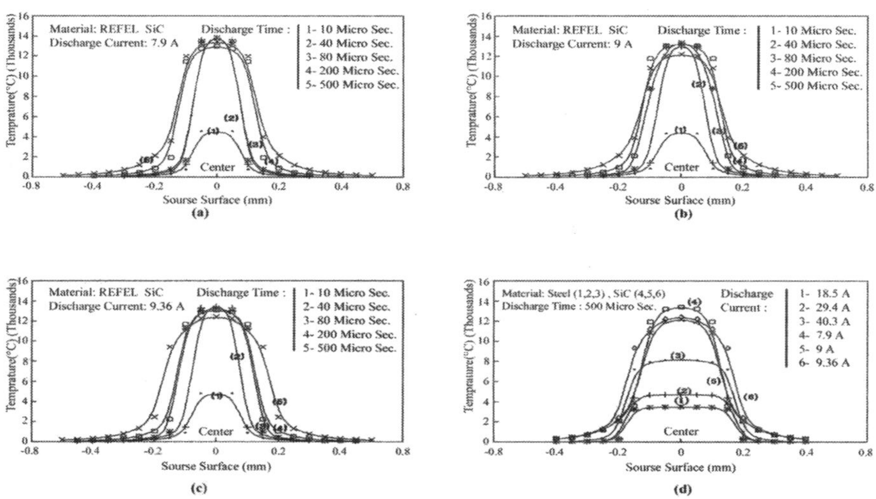

Figure 4: Variations of workpiece surface temperature with depth under various pulse durations and currents.

6 Conclusions

In electro-discharge machining of SiC, the voltage drop in workpiece is very high and the same as discharge voltage. For this high voltage reduction an important part of energy in the workpiece body is wasted as Joule Heating. This energy loss reaches its maximum on the workpiece surface for short pulse durations, and near the surface in depth for long pulse durations. The temperature of workpiece is highest in the centre of melting zone on the workpiece surface and in all directions, it reduces rapidly. In steel, the maximum temperature is lower and its reduction in depth and surface occurs in shorter distance. An increase in the discharge currents causes more voltage dropped in

SiC and a little increase in Joule heating. In comparison with SiC, the heating zone in steel is larger and with increasing in the discharge currents, this area increases extremely.

References

[1] Brook, R.J., Concise Encyclopaedia of Advanced Ceramic Materials, Max-plank-Institut für Metallforschung Stuttgart, Germany, 1991.
[2] Kennedy, p. & Shenan, J.V., Engineering Applications of REFEL Silicon-Carbide, Proc. of 3rd Conf. on Silicon-Carbide; Miami, USA, 1973.
[3] Reid, D.S. & Shaw, J.W., The machining of SiC Reactor Fuel Element Laboratories, UKEKA, Springfield, Preston, April 1969.
[4] Humphrey, P.D., An investigation into the electrical discharge machining of SiC, Reactor Fuel Element Laboratories, UKAEA, Springfield, Preston, 1976.
[5] Forrest, C.W., Kennedy, p. & Shenan, J.V. , The fabrication and properties of Self-Bonded SiC bodies, Special Ceramics, British Ceramic Research Association, June 1972.
[6] Tsai, K. M. & Wang, P. J., Semi-empirical model of surface finish on electrical discharge machining, International Journal of Machine Tools & Manufacture, Vol.4, No.10, pp. 1171-1179, August 2001.
[7] Yadav, V., Jain, V.K. & Dixit, P.M., Thermal stresses due to electrical discharge machining, International Journal of Machine Tools & Manufacture, Vol.42, No.8, pp. 877-888, June 2002.
[8] Kulkarni, A., Sharan, R. & Lal, G.K., An experimental study of discharge mechanism in electrochemical discharge machining, International Journal of Machine Tools & Manufacture, Vol.42, No.10, pp. 1121-1127, August 2002.
[9] Van Dijck, F., Physico-Mathematical Analysis of the Electro Discharge Machining Process, PhD Thesis, Catholic-University of Leuven, 1973 (English translation).
[10] Van Dijk, F., Crookall, J.R., Heuvelman, C.J., & Snoeys, R., Some Results of Physical Research in EDM, Proc. of the 5th Int. Symposium for Electro machining, pp. 68-85, 1977.

Section 4
Modelling of coatings

Numerical 3D BEM simulation of the chromium layer thickness distribution on parts in a rack plating configuration

J. Deconinck[1], G. Floridor[1], B. Van den Bossche[2], L. Bortels[2] & G. Nelissen[2]
[1]*Vrije Universiteit Brussel, Department of Electrical Engineering, Belgium*
[2]*ElSyCa N.V., Zellik, Belgium*

Abstract

A three dimensional Boundary Element Method (BEM) approach is used to compute the current density and plated layer thickness distribution on parts of complex shape in a rack plating configuration. Results are shown for a decorative chromium plating process. Simulations take into account electrode polarisation characteristics and the plating efficiency. The entire plating tank configuration is considered, including dimensions and position of anode baskets or rods, the number of parts on the rack, and the position of the rack in the tank. For decorative chromium plating, simulated results reveal that parts located at the outer positions of the rack receive a current that might double the value received by a central part, hence also the average plated thickness from one part to another might differ by a factor 2. The variation of the plated thickness on a single part is up to a factor 5 or more, with some of the inner zones receiving no layer thickness at all. These results are in agreement with industrial experience. The BEM solver is integrated in the SolidWorks CAD environment, enabling a fast modification of the dimensions of the entire configuration (rack configuration, current thieves, etc.), the surface mesh characteristics (determining the computational accuracy), and the physico-chemical input parameters (polarisation and efficiency characteristics, electrolyte conductivity, imposed DC or bipolar currents, etc.).
Keywords: electrodeposition, plating, advanced software packages, BEM.

1 Introduction

Although advanced numerical methods exist and have been applied with great success in a broad range of engineering domains (hydrodynamics, aerodynamics, structural mechanics, heat transfer, etc...), the use of these methods for electrochemical plating applications remains rather limited. One of the main reasons is the complexity of the physical, chemical and electrical phenomena governing electrochemical reactor behavior. In general, a complex interplay of the following phenomena takes place: electrochemical electrode kinetics, electrolyte hydrodynamics, ionic mass transport, gas evolution, and heat generation in the bulk and at the electrode-electrolyte interfaces.

From an engineering point of view however, the main focus is on the current density and layer thickness distribution, which principally depend upon the following phenomena:

- ohmic drop in the electrolyte solution;
- cathodic polarization and plating efficiency;
- anodic polarization;
- plating tank configuration including anode positioning, screens and current thieves;
- work piece shape and dimensions;
- selective insulation of work piece surfaces;
- number and position of work pieces on a rack;
- total current injected and anode and work piece contacting method.

The simplest modeling approach that takes into account these phenomena is commonly denoted as the 'potential model'[1]. In order to produce reliable simulation results, the physico-chemical input parameters (polarization behavior, plating efficiency and electrolyte conductivity) need to be defined carefully for the electrolyte bath being used, at a given operating temperature.

In this work, the ElSyCa PlatingMaster v1.0 software tool [2] is used to design a dedicated reactor for the selective chromium plating of an oil pump shaft. The current density and layer thickness distribution are evaluated for each step in the reactor design process. In a first step, the work piece is partly insulated (taped), in order to redirect the current to the surfaces that will be subject to wear and corrosion in practical mechanical operation conditions of the pump. A tubular reactor with cylindrical anode mesh is used. In the next steps, the anodes are redesigned, abandoning the concept of a single far away anode mesh for multiple anode meshes closer to the targeted surfaces (though still connected to one single current source). The anode positions and dimensions are optimized until a layer thickness distribution is obtained meeting certain minimum and maximum specifications. Plating within these product specifications avoids performing (expensive) additional grinding operations.

The ElSyCa PlatingMaster tool is entirely integrated in the SolidWorks® CAD environment hence enabling a very fast adaptation of the reactor design at each step. The unique combination of a user-friendly and powerful CAD tool with a fast and accurate current density and layer thickness distribution simulation tool allows to design and optimize the entire reactor within a couple of hours. The ElSyCa PlatingMaster tool can also be used for the evaluation of current density and layer thickness distributions on electrodes with high internal resistivity e.g. wafers[3], or for electrochemical machining applications [4,5].

2 Mathematical model and solution method

A brief overview of the mathematical model behind the current density and layer thickness distribution simulations is given below [1].

Although cathodic deposition reaction mechanisms can become very complex, the polarisation behaviour for single metal deposition processes $Me^{z+} + ze^{-} \rightarrow Me$ is often quite accurately described by a Butler-Volmer type relation:

$$j_n = j_0 \cdot \left(e^{\alpha_a F / RT(V-U)} - e^{-\alpha_c F / RT(V-U)} \right) \tag{1}$$

where j_n is the amplitude of current density normal to the electrode surface, j_o is the exchange current density, R the gas constant, T the temperature, α_a and α_c the anodic and cathodic charge transfer coefficients, and E_{01} the equilibrium potential for the deposition reaction. V and U hold for the electrode respectively electrolyte potential.

The main reaction occurring at inert impressed current anodes is oxygen evolution. Due to the thin passivation layer that is often present on the surface of this type of electrodes, a linear relation can describe the polarisation behavior:

$$j_n = A(V - U) + B \tag{2}$$

with A and B polarisation constants. This approach is sufficiently accurate.

The ohmic drop effects in the electrolyte are described by the Laplace equation for the electrolyte potential U:

$$\overline{\nabla}(\overline{j}) = 0 \qquad \overline{j} = -\sigma\overline{\nabla}U \tag{3}$$

On insulating boundaries, the current density perpendicular to the surface should be zero, which results in the following boundary condition:

$$\overline{j}.\overline{1}_n = j_n = -\sigma\overline{\nabla}U.\overline{1}_n = 0. \tag{4}$$

On electrodes, j_n is given by equations of type (1) or (2).

The local metal thickness Δd on the cathode is simply computed from Faraday's law:

$$\Delta d = \theta \frac{M\, \Delta t\, j_n}{\rho\, z\, F} \tag{5}$$

where θ is the efficiency for the deposition process (depending on the current density j_n). M holds for the atomic weight of the metal, ρ for the density, z is the number of electrons exchanged in the metal deposition reaction, and F is Faraday's constant.

The physico-chemical input data (polarisation behavior, plating efficiency and electrolyte conductivity) as used in this paper are those for a typical industrial hexavalent Cr bath.

2.1 Numerical solution method

As the conductivity of the computational domain is constant, BEM is to be preferred over other discretisation methods as Finite Element (FEM), Finite Volume (FVM) and Finite Difference Methods (FDM), to solve the simplified charge conservation equation (5). When the BEM is applied, only the boundaries of the domain are to be discretised. Another advantage, particularly for electrochemical systems, is that the current density distribution along the electrodes is a direct unknown to this problem, rather than a variable that has to be computed afterwards from the derivative of the potential field perpendicularly to the electrodes. This turns BEM into a more accurate method for potential problems, compared to FEM, FVM and FDM.

The characteristic BEM equation for the contribution to node i is [7]:

$$c^i U^i + \int_\Gamma U \frac{\partial w^*}{\partial n} d\Gamma = \int_\Gamma w^* Q \, d\Gamma \tag{6}$$

with

$$w^* = \frac{1}{4\pi r} \tag{7}$$

the 3D Green function ($\bar{r}$ representing the position vector) and $Q = \frac{\partial U}{\partial n}$ the flux at the boundary nodes. c^i is the space angle in node i. Γ is the 2D surface that encloses the 3D computational domain.

In order to apply BEM, the boundary Γ is to be discretised into a series of N non-overlapping elements, transforming equation (7) into:

$$c^i U^i + \sum_{j=1}^{N} \int_{\Gamma_j} U \frac{\partial w^*}{\partial n} d\Gamma = \sum_{j=1}^{N} \int_{\Gamma_j} w^* Q \, d\Gamma \tag{8}$$

The index j ranges over all elements of the domain and integration is performed over the surface Γ_j of each element.

Triangular elements with linear shape functions for the unknown potential U and current density Q are used for the 3D BEM computations as presented in this paper, restricting the unknowns to the N nodal values.

Taking into account equation (7), the BEM system of equations can be expressed in matrix form:

$$[H] \cdot \{U\} = [G] \cdot \{J_n\} \tag{9}$$

with $\{U\}$ and $\{J_n\}$ unknown vectors of size N. Once the boundary conditions have been introduced, i.e. zero current on insulators and overpotential relation (2) on electrodes, the system takes the form:

$$[H]\cdot\{U\}=[G]\cdot\begin{Bmatrix} J_n = f(V-U) \\ J_n = 0 \end{Bmatrix} \tag{10}$$

The matrices H and G in equation (10) are fully populated and depend only on the geometry of the system and the conductivity of the domain. This system of equations can be rewritten to group all unknowns in the left hand side vector:

$$[H^*]\cdot\begin{Bmatrix} J_n \\ U \end{Bmatrix}=[G^*]\cdot\begin{Bmatrix} V-\eta(j_n) \\ 0 \end{Bmatrix} \tag{11}$$

In most practical cases, system (11) is non-linear, except when $\eta(j_n)$ is linear in j_n.

The non-linear system equation matrix is solved using a Newton-Raphson iterative method, combined with a standard Gauss algorithm to tackle the linear system of equations that appears after each iteration of the Newton-Raphson procedure.

The electrode polarization data can be specified either as a linear or an exponential (Butler-Volmer) expression. Also a numerical data set can be used, in this case the data couples $(V\text{-}U, j_n)$ are transferred into a continuous derivable curve using a spline polynomial function.

An automated grid generator is used to produce the surface mesh that is required for the BEM computations[7]. This flexible grid generator creates an unstructured or structured mesh for each surface separately, or an unstructured mesh with structured boundary zones. The construction of a mesh that is optimally refined towards zones where edge effects are expected, and that generates coarse elements at any other position, allows to produce reliable numerical results with a restricted number of elements, hence limiting the computational efforts to solve the system of equations to a minimum.

3 Application

3.1 Configuration

The configuration consists of a brick shaped plating tank containing a rack with 20 parts (covers of an oil pump) as shown in figure 1. The anodes baskets (not shown) are positioned on the front plane. The biggest challenge in rack plating is ensuring uniform plating over the entire rack. Due to the edge effects, the parts located on the outside of the rack attract more current. This is further complicated by the fact that in most applications, the parts are placed manually on the fixtures, thus leading to sometimes badly positioned parts on the rack. To simulate the influence of the placement on the current density and deposit distribution, one of the parts is protruding (closer to the anodes) the plane of the other parts and one part is positioned further away from the anodes. Because the potential model assumes the mass transport to be ideal, the flow of the electrolyte does not need to be determined.

The entire configuration was defined in the SolidWorks© CAD package, and simulations were performed with the ElSyCa PlatingMaster© v1.0 software tool, which is entirely integrated in this CAD package[2].

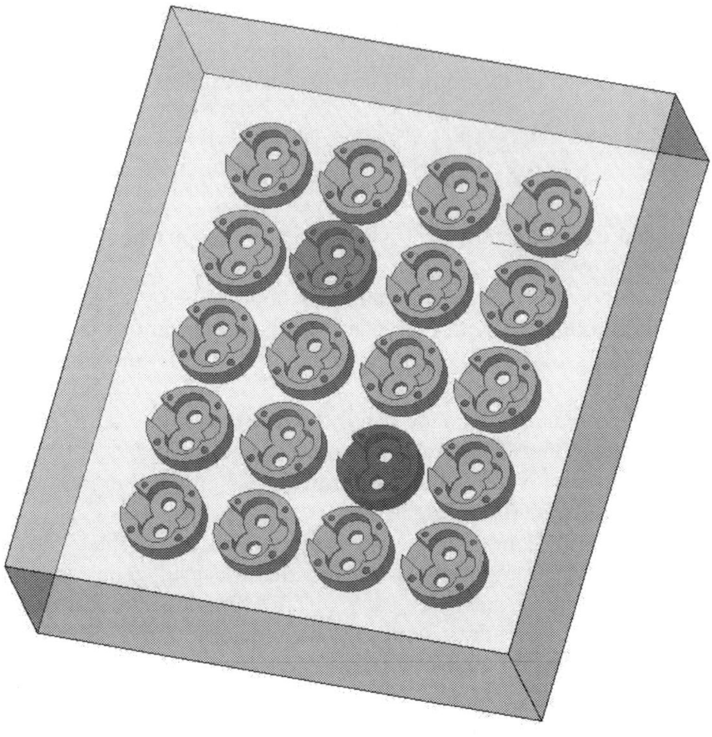

Figure 1: Configuration of 20 parts on a rack, with the red part (2, 2) protruding and the blue part (3, 4) moved to the back.

3.2 Boundary conditions

A total current of 700 A is imposed on the anode baskets, leading to a mean current of 35 A per part. In its turn, this gives an average current density of about 5000 A/m^2 on the parts. The total plating time is set to 10 minutes.

3.3 Surface mesh

A rather fine triangular non-uniform surface mesh is used for the tank oil pump as shown in figure 2. The outer walls of the plating tank and the anodes (not shown) are discretized using larger triangular elements. This ensures an optimal trade off between computational efforts and accuracy of the simulated results.

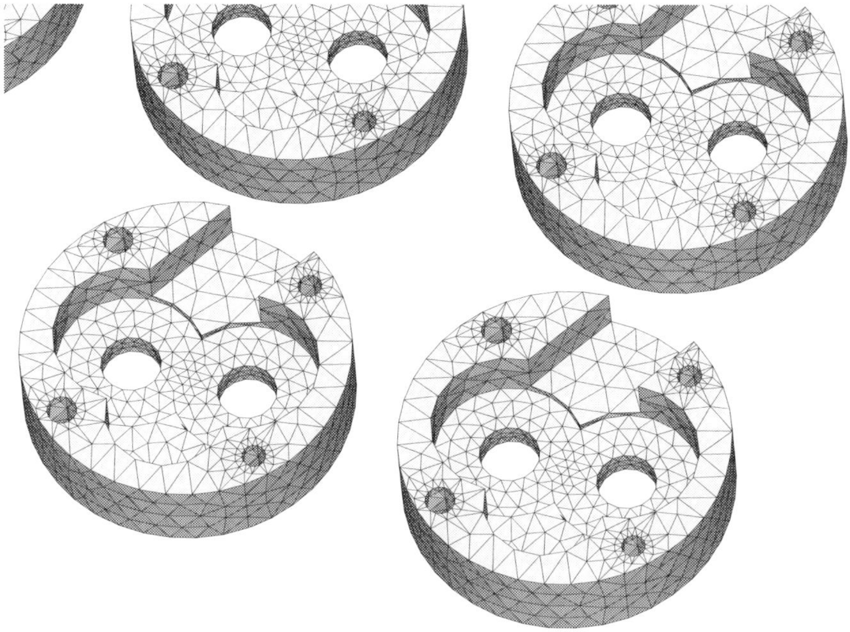

Figure 2: Triangular surface mesh for the oil pumps.

3.4 Computational specifications

The total number of elements used to solve the BEM problem is 57196 triangles and 38452 points.

The CPU time is about 30 minutes on a 1400 MHz Pentium IV PC for 7 iterations in the Newton-Raphson procedure, after which a sufficient convergence level of 1E-5 is reached.

3.5 Results and discussion

Figure 3 shows the resulting deposit thickness (in μm) distribution over the entire rack. As expected the parts on the outer side of the rack receive a higher current density. As seen in figure 3, these edge effects can easily lead to a deposit for a part positioned towards the outer rim of the rack (10 μm or higher) that is double the deposit of a part in the middle of the rack (about 5 μm).

Also the part positioned closer to the anodes has a higher deposit. On the other hand, the part placed further away from the anode does not get plated at all. From this one can conclude that an accurate positioning of the part on the fixtures is a very critical issue when trying to achieve uniform chromium plating thickness in rack plating applications.

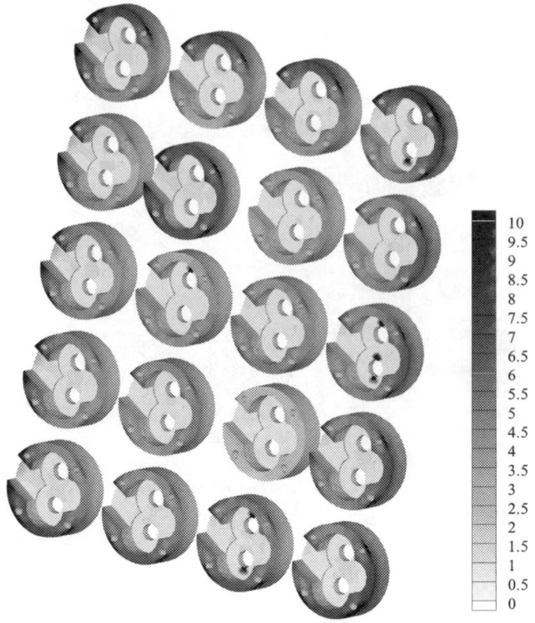

Figure 3: Deposit thickness distribution in μm over the rack.

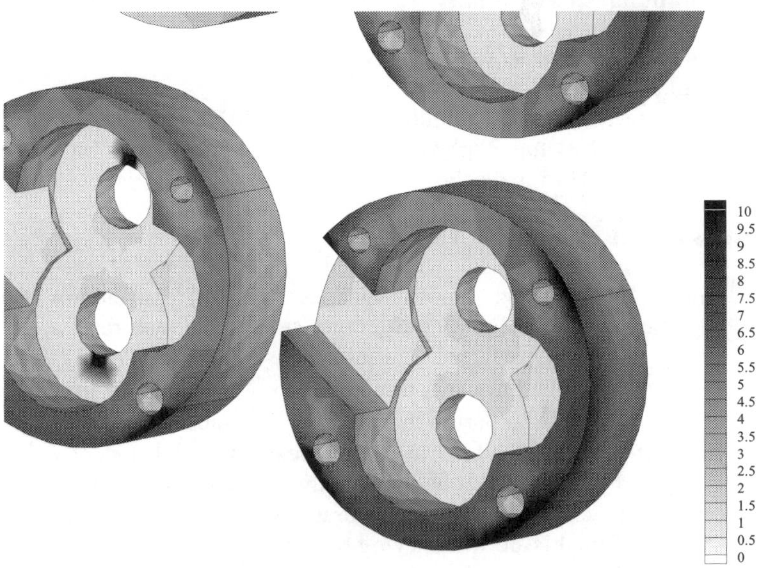

Figure 4: Deposit thickness distribution in μm over a single part.

The variation of the deposit thickness over a single oil pump is shown in figure 4. The interior receives much less chromium than the outer edges. The variation in plating thickness over a single pump varies from almost zero to more than 10 μm at the sharp edges. It is also clear that if a very uniform deposit over each part is required, additional changes to the plating configuration are needed. This can include the use of shielding, auxiliary anodes and/or current robbers.

4 Conclusions

The capabilities of the ElSyCa PlatingMaster tool for the fast and accurate analysis of a rack plating application have been illustrated. From the calculations is becomes clear that a thorough design and optimisation of a rack plating configuration is required possible.

Finally, the ElSyCa PlatingMaster software enables manufacturers to discuss the manufacturability of a product (cost, quality) based upon sound analytical data with their customers of Design and/or Engineering departments.

References

[1] J. Deconinck, G. Maggetto, J. Vereecken, J. Electrochem. Soc., 132 (1985) 2960.

[2] ElSyCa PlatingMaster v1.0 manual, www.elsyca.com.

[3] M. Purcar, B. Van den Bossche, L. Bortels, J. Deconinck, G. Nelissen, J. Electrochem. Soc. 151-9 (2004) D78-D86.

[4] L. Bortels, M. Purcar, B. Van den Bossche, J. Deconinck, J. Mater. Process. Tech. 149 1-3 (2004) 472-478.

[5] M. Purcar, L. Bortels, B. Van den Bossche, J. Deconinck, J. Mater. Process. Tech. 149 1-3 (2004) 486-492.

[6] C.A. Brebbia, Boundary Element Techniques – Theory and Applications in Engineering, Springer-Verlag Berlin, Heidelberg, 1983.

[7] Athanasiadis A.N., Deconinck H., Int. J. Numer. Meth. Eng. 58 (2003) 301.

BEM analysis of the protective mechanism of patinas against atmospheric corrosion

A. Peratta[1] & E. Ayllón[2]
[1]Wessex Institute of Technology, Southampton, U.K.
CITEFA, Instituto de Investigaciones Científicas y Técnicas de las
[2]Fuerzas Armadas, BsAs, Argentina

Abstract

Weathering steels are characterised by a low content of alloying elements such as Cu, Cr, Ni, P, and Si, which contribute to the outdoor formation of a corrosion product adherent film called a patina. Its attractive appearance and passivity power towards the environment to which it is exposed makes paint or other coatings for initial protection and subsequent maintenance unnecessary. Since its introduction in 1933, these attributes have spread its use to a large variety of structural applications such as buildings, bridges and electric power transmission line towers. Patinas are made up of compounds of α, β, γ, and δ iron oxyhydroxides (FeOOH) and magnetite in different proportions. Their composition, density and morphology depend on the environmental conditions of the place they are formed in. At present, no model explaining patina behaviour as an effective protection of steel against atmospheric corrosion has gained general acceptance among researchers. The purpose of this work is to propose a porous media model that shows the influence of patinas compactness as a protector. The initial stages of the attack during exposure without outdoor protection is in fact a selective one, being the non-metallic inclusions the most remarkable points of corrosion. The controlling mechanism of the patina protective attribute results from the compactness provided by the microalloying elements which leads to a decrease of the pores and particle size. The transport model of Fe^{2+} through the patina for weathering steels was analysed by a 3D computational code based on the Boundary Element Method for solute transport in porous media, and compared to the transport through non-weathering steel, thus relating the electrochemical nature of the corrosion to the porosity of the patina.

WIT Transactions on Engineering Sciences, Vol 48, © 2005 WIT Press
www.witpress.com, ISSN 1743-3533 (on-line)

1 Introduction

Since its introduction in 1933, the concept of patinas as corrosion controlling mechanism has been playing an important roll in many structural applications such as buildings, bridges and electric power transmission line towers. Patinas are made up of compounds of α, β, γ, and δ iron oxyhydroxides (FeOOH) and magnetite in different proportions. Their composition, density and morphology depend on the environmental conditions of the place they are formed in. At present, no model explaining patinas behaviour as an effective protection of steel against atmospheric corrosion has gained general acceptance among researchers.

The purpose of this work is to propose a porous media (PM) model that shows the influence of patinas compactness as a protector. Since the beginning of the systematic study of atmospheric corrosion, this type of corrosion was considered basically an electrochemical phenomenon. The analysis of the initial stages of the corrosive attack showed that this phenomenon which was considered uniform at first- is actually a moderate one being the non-metallic inclusions in structural steels the preferred points of attack. Evan's electrochemical model for the atmospheric corrosion shows the reactions that explain the behaviour of corrosion products films as well as the influence of microalloying elements on the size of colloidal aggregates particles that form patinas. There have been many proposals to explain the protection mechanism. However, there seems to be no generally accepted theory up to now [1]. In the beginning of the research work there were two tendencies: the first one was to attribute the protective nature to the chemical characteristics of the present species, and the second one was to attribute the said protective nature to the morphology of the protective films. Many techniques were used for the patina characterisation of corrosion products, which were formed because of being exposed to a natural environment or in a laboratory. Non-metallic inclusions are the most abundant ones, MnS inclusions with or without Fe have a main role at the beginning of the corrosion process. G. Wrangler established that some of these inclusions are "active", and others are "inactive" (ie, they do not take part in the whole process). According to the author, around the MnS inclusions there is an area (well determined by E. Klund [2]) attributed to the steel cooling rate from its tapping temperature, following the inclusion contour. This area is called MnS micro inclusions infected area, and is very reactive. While the MnS inclusion works as a cathodic area, the said infected area works as an anodic one. The progress of this process leaves the non-metallic inclusion without support to the steel matrix, and it is removed. Therefore, a whole or pit filled with corrosion products is originated and acts as an anodic area. Sometimes it is the inclusion itself the one that dissolves. This occurs when the inclusion has other chemical elements, which are easily attacked like the lanthanum.

2 Evan's electrochemical model

Outdoor tests carried out during long periods of exposure have revealed that weathering steels (WS) with small additions of Cu, Cr, and P have much better atmospheric corrosion resistance than those structural steels without microalloying elements. Laboratory research by M. Poubaix *et al.* [3] [4] has clearly shown that the said difference only occurs when steel samples are subject to repeated cycles of drying out and humidification. These tests like the ones carried out in a natural environment with a relative humidity always near saturation [5] have shown that there is no advantage in substituting ordinary steels for WS under the said environmental conditions. Many authors [6] [7] have pointed out the key influence of Cu as the most significant microalloying element to enhance atmospheric corrosion resistance, and they relate Cu to the formation of an inner denser and tightly adherent layer, which coats steel without cracks or defects. According to this, the tests carried out by Electron Probe Microanalysis (EPMA) [8] show that Cu concentrates mainly in those cavities resulting from the initial attack to non-metallic inclusions, and in the space among them, which is in contact with steel. The intensity of the signal recorded in those tests is proportional to the Cu amount, and increases according to the outdoor exposure time, showing also an increase in the protective nature of the patina from 1 to 3 years. Infrared (IR) spectroscopy tests show that the main components of protective patinas are first the γ ferric oxyhydroxides (γ-$FeOOH$), called lepidocrocite, and then the α and β oxyhydroxides and magnetite (Fe_3O_4) in smaller amounts. The atmospheric corrosion product films are secondary aggregates of colloidal primary particles. The aggregation condition increases with the ageing of the film and Cu aggregates. It has been reported that this element inhibits the increase of colloidal particles modifying their protective properties [9]. Furthermore, K. Inouye *et al.* [10] carried out a series of lab tests with colorimetrically determined $\gamma FeOOH$ with $CuSO_4$ to get solutions with Cu contents from zero to a Cu/Fe ratio of 50%. The results revealed the growing influence of the Cu with respect to the Fe, showing a reduction in the size of the colloidal particles, which in a Cu/Fe ratio of 20% is inferior to $0.05/0, 1\mu m$. The protective features of the corrosion product films would be then associated to their compactness [11]. The reduction of the particle size resulting from an enrichment of Cu occurs with a reduction of the pores size and a greater tortuosity in the aggregate. According to Evan's model, the reduction conditions exist in a big part of the film if the pores are filled with water (from rain or mist). However, the oxidation conditions prevail during the drying out periods. The balance between these two conditions moves to the steel during the drying out periods and to the atmosphere during the humidification periods. Therefore, there are areas of the film that would be reduced by the following sequence of reactions, where the Fe^{+2}_{aq} shows the Fe^{+2} ions passing to the aqueous solution (electrolyte): $Fe \rightarrow Fe^{2+}_{aq.} + 2e$, and $2\gamma FeOOH + 2H^{+}2e \rightarrow 2Fe\left[OHOH\right]$ in the $\gamma FeOOH$ crystal surface, which under reduction conditions near the steel surface may be: $4\gamma FeOOH + 2Fe\left[OHOH\right] \rightarrow 2Fe_3O_4 + 4H_2O$, and $Fe^{2+}_{aq} + 2\gamma FeOOH \rightarrow Fe_3O_4 + 2H^{+}$. If instead of the $\gamma FeOOH$ reduction, O_2 is benefited from the drying out conditions of the film pores, the reoxidation conditions of the intermediary Fe^{+2} and Fe_3O_4 change. And the following reactions may occur: $2Fe\left[OH.OH\right] + 0.5O_2 \rightarrow 2\gamma FeOOH + H_2O$ and $2Fe_3O_4 + 0.5O_2 \rightarrow 3\gamma Fe_2O_3$. In the following cycle, γFe_2O_3 may be again reduced to Fe_3O_4. The electrons required for this process are produced by the anodic dissolution of steel: $Fe \rightarrow Fe^{2+} + 2e$, and $4\gamma Fe_2O_3 + Fe^{2+} + 2e \rightarrow 3Fe_3O_4$.

Therefore, the intermediary Fe^{+2}, the Fe_3O_4 -reduction product-, the γ-$FeOOH$ and the γ-Fe_2O_3 -oxidation products- become the constituents of the redox process of the corrosion product film. The reaction $Fe \rightarrow Fe^{+2}_{aq} + 2e^-$ is produced in the anodic cavities or sulphate nests. The electrons must be consumed by the supplemented cathodic reaction in the area among anodic cavities, reducing the superficial layer of the $FeOOH$ oxyhydroxide globules, as follows: $FeOOH + e^- + H^+ \leftrightarrows [FeOHOH]$. In this reaction, the H^+ is consumed, alkalinising the solution in the pores. The Fe^{+2} ion may cross the solid-electrolyte interface as follows: $[Fe.OH.OH] + 2H^+ \rightarrow Fe^{+2}_{aq} + 2H_2O$.

According to the results reported by M. Stratmann *et al.* [12], the latter is particularly in the case of a pH between 4.3 and 5, as it has been experimentally established for acid rains [13]; and it seems not to be relevant for a $pH \geq 6$. The localised alkalinity reduces the rate of the said reactions, and therefore it dissolves the $FeOHOH$ superficial film, with a passage of Fe^{+2} to the electrolyte. The excellent atmospheric corrosion properties of WS cannot entirely be explained by the alloying elements. Although it seems, that the phosphorus in steel improves the atmospheric corrosion resistance, it seems also, that sulphur or sulphur oxides in air have an effect on protective patina formation on the WS. This can probably be explained by the means of stopped sulphuric acid regeneration cycle, [14] Under these localised alkalinity conditions, the resulting stable phase is the magnetite (Fe_3O_4), as shown in the pH - potential diagrams by Pourbaix [15], and Suzuki *et al.* [7]. Then, the cathodic reaction would be as follows: $4FeOOH + 2\,[FeOHOH] \leftrightarrows 2Fe_3O_4 + 4H_2O$ and also: $Fe^{+2}_{aq} + 2FeOOH \leftrightarrows Fe_3O_4 + 2H^+$. This shows that the Fe^{+2}_{aq} contribution is important for magnetite formation. Electrochemical curves show that the potential at which the maximum anodic current is produced, the oxidised element is now the Fe^{+2} intermediary in the surface of $FeOOH$ globules instead of the magnetite [12]. Anyway, this process depends on the time and/or electrochemical potential at which the cathodic reduction is performed: $2\,[Fe.OH.OH] + 1/2O_2 \rightarrow 2FeOOH + H_2O$. The $FeOOH$ may reform as a δ-$FeOOH$ compound due to the microalloying elements resulting from the anodic reaction of steel dissolution $2[Fe.OH.OH] + 1/2O_2 \rightarrow 2\delta FeOOH + H_2O$. Moreover, it would protect the patina inner layer due to an increase of its compactness, according to what is mentioned in the tests by E. Inouye *et al.* [10, 11, 8]. Where there are neither microalloying elements or long humidification periods the $\delta - FeOOH$ would have been transformed into Fe^{O4}_3, because there is not controlling effect on the Fe3O4 production, which does not have a protective attribute, as reported by H. Kihira *et al.* [16].

3 Porosity influence

The porosity influence on patinas behaviour of the corrosion products with respect to the atmospheric corrosion was tested by a PM model. The Δp thickness loss of the samples exposed outdoors is the result of a uniform attack. However, this is only the macroscopic side of the problem. A SEM analysis of the first stages of the corrosive attack showed that in those places initially filled by non-metallic inclusions, there appeared the so-called anodic cavities or sulphate nests with semi-spherical morphology. If we suppose that all the mass lost during the corrosive process comes from the anodic cavities volume, then Δp can be estimated according to $\Delta p = \pi n d^3/12$, from where the mean diameter d of the cavity becomes:

$$d = \left(\frac{12\Delta p}{\pi n}\right)^{1/3}, \tag{1}$$

where n is the number of non-metallic inclusions per area exposed to the attack. The ions Fe^{+2} coming from the sulphate nests are released to the patina, regarded as continuum PM. Hence, the transport of Fe^{+2} throughout it can be described by an advection-diffusion model with steady state (incompressible) flow conditions driven by external bulk effects. Thus, the concentration (c) of Fe^{+2} ($c[Fe]$)obeys the following differential equation:

$$\phi\frac{\partial c}{\partial t} + \nabla \cdot \mathrm{q} = -\lambda c, \tag{2}$$

where ϕ is the porosity of the medium, t represents time, $\mathrm{q} = \mathrm{v}c - D\nabla c$, represents the solute flux and D is the hydrodynamic dispersion tensor given by:

$$D_{ij} = \phi\delta_{ij}\mathcal{D} + \alpha_L P_{ij}(\mathrm{v}) + \alpha_T\left(\delta_{ij} - P_ij\right)\mathrm{v} \tag{3}$$

where α_L and α_T are the longitudinal and transversal dispersion coefficients, respectively; $\mathcal{D}$ is the molecular diffusivity, P_{ij} is the projector onto the velocity vector v, and δ_{ij} is the Kronecker delta. The flow is governed by Darcy law and the mass conservation equation in which $\mathrm{v} = -k\nabla h$, being v the average velocity over the representative volume in the sense of ref. [17], h the hydraulic head and k the hydraulic conductivity. (v is normally lower than the infiltration velocity, particularly when tortuosity is high.) Therefore, regarding steady state conditions the hydraulic head obeys the potential equation: $\nabla\left(-k\nabla h\right) = 0$. Adaptation of the Blake-Kozeny relation for PM in case of spherical particles yields the following expression for k:

$$k = \frac{\rho g}{\mu}\frac{D_p^2}{150}\frac{\phi^3}{\left(1-\phi\right)^2}, \tag{4}$$

where D_p is the particle diameter, ρ is the density of the solution and μ is the viscosity [18, 19, 20]. Analysis with SEM in WS led to the conclusion that $D_p \approx 0.05\mu m$ in the inner layer in contact with the steel and increases up to $D_p \approx 10\mu m$ towards the outer surface.

4 Numerical modelling

Numerical modelling of flow and transient solute transport in PM by Boundary Element Method (BEM) has been successfully implemented in the last decade by many researchers [21, 22, 23, 24]. The main advantage of the formulation is that volume meshes can be avoided, and the accuracy can be improved since it employs the fundamental solution of the leading operator of the partial differential equation. In this work, a three dimensional multi-domain implementation of the Dual Reciprocity BEM for the time dependent advection diffusion equation has been employed (DRM-MD-BEM). This BEM is based on the fundamental solution of the laplace operator, any other term apart from the diffusive one is integrated by the dual reciprocity method; a thorough explanation of the formulation can be found in [25]. The integration domain is decomposed into many subdomains (tetrahedrons) with the aid of a customisable finite element type preprocessor and mesh generator [26]. In the computational code, the flow problem is solved first with the standard BEM as described in [22] and then the velocity field calculated is introduced as an input to the transport problem.

5 Model definition

The conceptual model representing a microscopic portion of the affected region in WS is shown in Fig. 1, in which the patina of thickness $\Delta p = 100\mu m$ is composed of two layers of porous media: inner and outer, respectively. For clarity, point O_1 in Fig. 1 is located at the origin whereas point O_2 is located at coordinates $(x, y, z) = (840, 100, 440)\mu m$. The sulphate nests (anodic areas) enter to the model as boundary conditions providing a fixed mass flux of Fe^{2+} given by:

$$m = \frac{i_c W}{ZnF}, \tag{5}$$

where $Z = 2mole^-/molFe$ is the number of moles of e^- per mole of Fe exchanged in the reaction, $n \approx 150mm^{-2}$ is the number of anodic zones per unit area, $W = 55.846g/mol$ is the atomic weight of Fe, $F = 96500C/mole^-$ is Faraday's law constant, and i_c is the anodic current density. The latter was estimated with the corrosion computational code GAMMACRUNCH [27]. In case of WS $i_c \approx 2 \times 10^{-7} A/cm^2$ thus yielding a mass flux $m = 3.334 \times 10^{-4} \mu g/day$ whereas for non-weathering steels (NWS) $i_c \approx 4.5 \times 10^{-5} A/cm^2$ resulting in $m = 7.5 \times 10^{-2} \mu g/day$. The mean diameter of the anodic zone has been adopted to be $d = 80\mu m$ according to experimental findings with SEM samples [10]. The boundary conditions are specified in Table 1 and the transport properties for WS are summarised in Table 2. Patinas are not present in NWS, however they still present a non-self-protecting porous film of typical particle size $D_p \approx 10\mu m$ with transport properties similar to the outer layer of the patina in WS. If the sample is at an angle of 45°, a small advective term arises yielding a constant uniform velocity field $\mathrm{v} = -k \sin(45^o)\,\hat{x}$.

6 Results

The proposed model has been solved for W and NWS. Fig. 2 shows the $c[Fe]$ for WS at different time levels during the transient period of the process in a 2D analogous model based on the DR-BEM. The black streaks represent the anodic

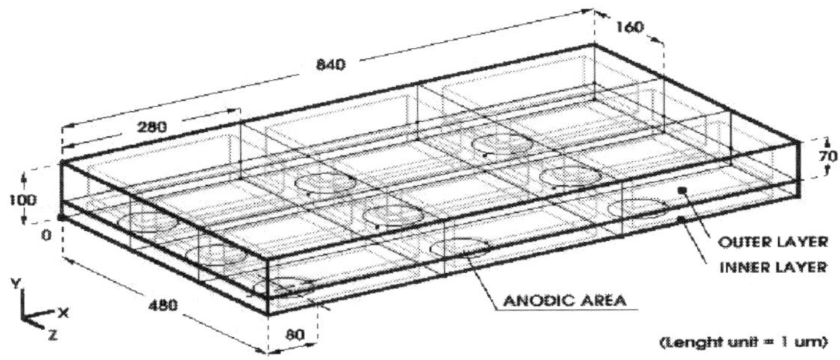

Figure 1: 3D conceptual model of the WS.

Table 1: Boundary conditions for conceptual models (WS and NWS).

Location	Flow	Transport
Plane yz at $x = 0$	$h = 5 \times 10^{-4}m$	$c = 0$
Plane yz at $x = 840\mu m$	$h = 0$	$q = 0$
Planes xy and yz (*)	$v_n = 0$	$q = 0$
Plane xz at $y = 0$ (anodic area)	$v_n = 0$	$q = q_{in}$

(*)Plane $x - y$ at $z = 0 \wedge 440\mu m$; plane $x - z$ at $y = 100\mu m$ and cathodic area at $y = 0$

Table 2: Physical properties of the test example (WS).

Name	Symbol	Unit	Inner	Outer
Thickness	Δp_i	μm	30	70
Porosity	ϕ_i	-	0.4	0.4
Hydraulic Conductivity	k_i	$\mu m/day$	1	400
Velocity	v_i	$\mu m/day$	$\sqrt{2}/2$	$200\sqrt{2}$
Molecular Diffusivity	$\mathcal{D}_i$	m^2/day	10^{-4}	10^{-4}
Longitudinal dispersivity	α_{Li}	μm	2	6
Transversal dispersivity	α_{Ti}	μm	1	3
Advection time scale(*)	τ_{Ai}	day	25	0.625
Diffusion time scale(**)	τ_{Di}	s	14	14

(*) $\tau_A = \Delta p/v$, where $\Delta p = 100\mu m$ is the characteristic length scale of the problem.
(**) $\tau_D = 4\phi\Delta p^2/D$, where $\Delta p = 100\mu m$ is the characteristic length scale of the problem.

zones for which the mass flux per inclusion $q_{in} = 3.334 \times 10^{-4}\mu g/day$ and $7.5 \times 10^{-2}\mu g/day$ have been prescribed for WS and NWS, respectively. For clarity, the upper limit of the colormap is different at each time level as shown at the bottom of the figure. The distribution of Fe^{+2} at $625s$ in the the 3D model of the patina of the WS can be observed in Fig. 3. The colour indicates $c[Fe]$ according to the colorbar on the right side of the figure. Fig. 4 shows a comparison between WS

Time $[s]$	20	40	60	500
C_{max} $[kg/m^3]$	0.12	0.21	0.29	1.00

Figure 2: Concentration field of Fe^{2+} in the patina for WS.

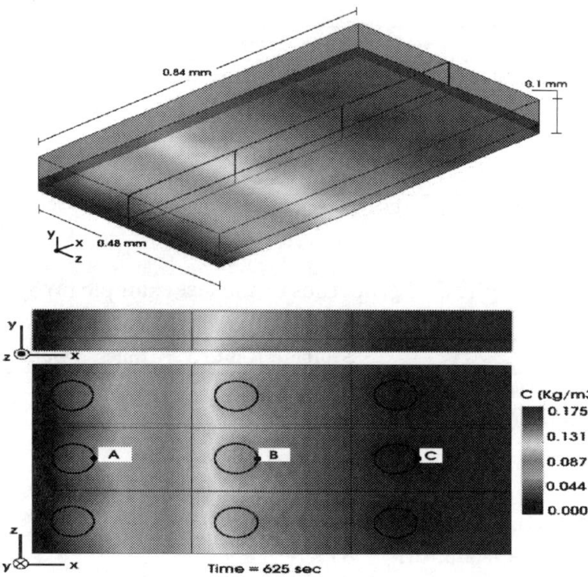

Figure 3: Concentration of Fe^{+2} at time $t = 625s$ in the WS. Observation points A,B, and C are located at the internal boundary of the corrosion patina, at coordinates $y = 0$, $z = 240\mu m$, $x_A = 120\mu m$, $x_B = 400\mu m$ and $x_C = 680\mu m$, respectively.

and NWS for the evolution of $c[Fe]$ at the observation points shown in Fig. 3. The transient span of both samples under saturated conditions is $\approx 500s$ and is mainly controlled by the porosity and the particle diameter as predicted by eq (4). The $c[Fe]$ in NWS is approximately 200 times larger in the patina of WS, as a direct consequence of the different corrosion currents prescribed at the anodic zones.

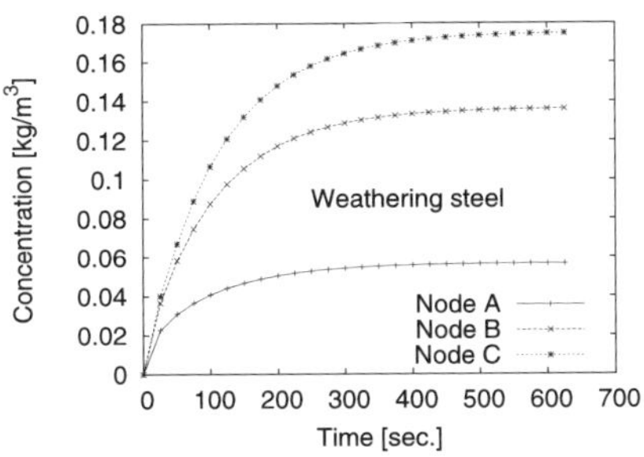

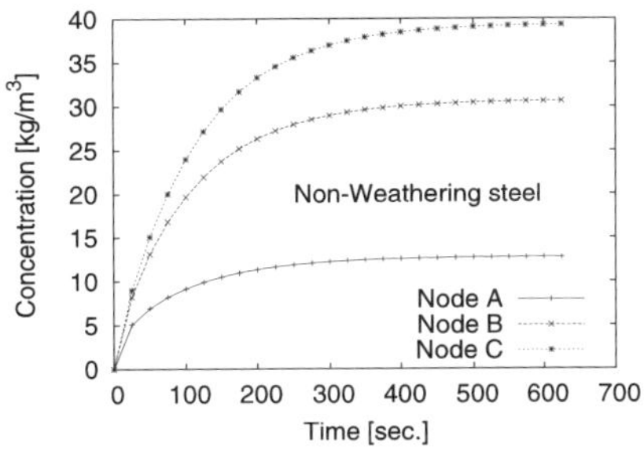

Figure 4: $c[Fe]$ at points A,B and C of Fig. 3 for WS (left) and NWS (right).

7 Conclusions

The controlling mechanism of the patina protective attribute results from the compactness provided by the microalloying elements, which leads to a decrease of the pores and particle size. The model can explain the effect of patina compactness on the ionic transport within it as a controlling factor of the reactions produced during corrosion and relates the electrochemical nature of the corrosion process to the porosity of the patinas. Transport of Fe^{2+} through the patina yields the formation of magnetite near the cathodic zone. Since magnetite is not self protective Kihira *et al.* [16] then it contributes to spoil the self protective nature of the patina. Concentration of Fe^{2+} in non-weathering steels achieves higher values in shorter time than in weathering steels due to the high values of current density and slow transport of the ions through the external layer.

Finally, porosity of the patina plays a key roll in controlling the transport of Fe^{+2} and the consequent formation of magnetite in weathering steels.

References

[1] Merino, P., Sanchez, A. & Porto, E., Mapas de españa de corrosividad atmosférica. *CYTED*, eds. M.M. Linares & S.F. Matas, p. 319, 1993.

[2] Klund, E. *Scandinavian Journal of Metallurgy*, p. 331, 1972.

[3] Pourbaix, M., *Corrosion Sci*, volume 12. p. 161, 1972.

[4] Pourbaix, M. & de Miranda, L., In passivity and its breakdown on iron and iron base alloys. *NACE*, eds. R.W. Sataehle & H. Okada, p. 47, 1976.

[5] Bohnenkamp, K., Tribune du CEBEDAU. Technical Report 324, 1970.

[6] T. Mizawa *etal*, *Corrosion Sci*, volume 11. p. 35, 1971.

[7] Suzuki, I., Masuko, N. & Hisamatsu, Y., *Corrosion Sci*, volume 19. p. 521, 1979.

[8] T. Moroishi, *Transaction ISIJ Sci*, volume 11. p. 390, 1971.

[9] Suzuki, J., Hitsamatsu, Y. & Masuko, N. *Journal Electrochem Soc Solid-State Science and Technology*, **127(10)**, p. 2210, 1980.

[10] K. Inouye et al, *Corrosion Sci*, volume 16. p. 507, 1976.

[11] Rosales, B.M., Superficial films behaviour during atmospheric corrosion. *Scientific Meeting on Fe Physicochemical and its Oxides*, INIFTA: La Plata, 1986.

[12] Stratmann, M., Bohnenkamp, K. & Engell, J., *Corrosion Sci*, volume 23. p. 966, 1983.

[13] Townsend, H.E., Simpson, T.C. & Johnson, G.L. *Proc. 12th International Congress on Metallic Corrosion*, Texas, USA, p. 624, 1993.

[14] Arpone, M. & Forsén, O. *14th International Corrosion Congress*, Cape Town, South Africa, 1999.

[15] Pourbaix, M., *Atlas of Electrochemical Equilibria*. Pergamon Press, 1966.

[16] Kihira, H., Ito, S. & Murata, T., *Corrosion Sci*, volume 45. p. 347, 1989.

[17] Bear, J., Tsang, C.F. & de Marsily, G., (eds.) *Flow and Contaminant Transport in Fractured Rock*. Academic Press, Inc.: San Diego, New York, Berkeley, Boston, London, Sydney and Tokyo, 1993.

[18] Geiger, G.H. & Poirier, D.R., *Transport Phenomena in Metallurgy*. Addison Wesley Publishing Co., p. 94, 1973.

[19] Hansen, D., Discussion of 'on the use of the kozeny-carman equation to predict the hydraulic conductivity of soils'. *Can Geotech*, **41**, pp. 990–993, 2004.

[20] Mauran, S., Rigaud, L. & Coudevylle, O., Application of the ck correlation to a high-porosity and anisotropic consolidated medium: The compressed expanded natural graphite. *Transport in Porous Media*, **43**, pp. 355–376, 2001.

[21] Peratta, A. & Popov, V., Numerical modeling of contaminant dispersion from underground mine repositories. *Waste Management 2004*, eds. C.A. Brebbia, A. Kungolos, V. Popov & H. Itoh, Wessex Institute of Technology, WIT Press: Rhodes, Greece, volume 00 of 2nd *Int. Conf. on Waste Management and Environment*, pp. 0–0, 2004.

[22] Peratta, A. & Popov, V., A new scheme for numerical modelling of flow and transport processes in 3d fractured porous media. Submitted to Advances in Water Resources in February 2004.

[23] Zagar, I. & Skerget, L., Integral formulations of a diffusive-convective transport equation. *BE applications in Fluid Dynamics*, eds. C.A. Brebbia & H. Power, Computational Mechanics Publications: Southampton Boston, pp. 153–176, 1995.

[24] Popov, V. & Power, H., The DRM-MD integral equation method: An efficient approach for the numerical solution of domain dominant problems. *Int J Num Meth Engng*, **44**, pp. 327–353, 1999.

[25] Peratta, A., *BEM applied to flow and transport in fractured porous media*. Ph.D. thesis, Wessex Institute of Technology, Southampton, UK, 2004.

[26] GID resources. Website. Http://gid.cimne.upc.es.

[27] Jensen, M. & Britz, D., *Corrosion Sci*, volume 21. p. 285, 1991.

Electrochemical evaluation of a high solid coating in acid media

M. Bethencourt, F. J. Botana, M. J. Cano, R. M. Osuna
& J. M. Sanchez-Amaya
Department of Materials Science, Metallurgical Engineering and Inorganic Chemistry, University of Cadiz, Spain

Abstract

In the present work a two-component solvent free amine cured epoxy coating was used to protect carbon steel. The coated metal was immersed in an aggressive acid solution and monitored by means of electrochemical techniques: Electrochemical Impedance Spectroscopy (EIS) and Electrochemical Noise (EN). These techniques were used to evaluate the coating degradation evolution. The values of the pore resistance R_{po}, the impedance at 0.1 Hz (calculated by means of both EIS and EN), and noise resistance (R_n) were the main parameters calculated with the aim of monitoring the evolution of the system along the immersion time. An excellent correlation among the EIS and EN data was found.
Keywords: EIS, ENM, coating, carbon steel, degradation, acidic media.

1 Introduction

Increasing the percentage of solids in conventional solventborne paints has represented one of the first attempts to decrease the volatile organic compounds (VOC) content of a coating formulation and comply with increasingly stringent environmental requirements [1]. Traditional high solid coatings use less paint to achieve the same thickness obtained with other conventional formulations. This results in the volatilization of less solvent into the environment.

By definition, a high solids coating typically contains greater than 60% solids by weigh or 80% solids by volume [2]. This technology solution is currently accepted by the automotive, aerospace, and marine industries for use in a variety of formulations. The major advantage associated with the use of this kind of coatings is that the same equipment can be used to apply new high solids and conventional solventborne paints. On the contrary, the major disadvantage with high solids paint lies in the higher viscosities exhibited by the coating formulations. In addition, although in less content, these coatings still contain a significant amount of traditional solvents and diluents [2].

Nowadays, epoxy solvent free coatings have been adopted in an important way by the marine industry. In this contribution, a two-component solvent free coating, based on epoxy resin and cured with amines, has been studied. This coating, which has excellent resistance to corrosion, chemical products and oil, is usually employed as a monolayer designed for the protection of steel structures and storage tanks.

In order to evaluate the anticorrosive protection of the coating in acid media, electrochemical measurements were recorded and interpreted. On the one hand, Electrochemical Impedance Spectroscopy (EIS) was used to analyse discriminately electrochemical response of the metal-paint system. On the other hand, Electrochemical Noise Measurements (ENM) completed the evaluation. Finally, correlation among all the measurements was intended.

2 Experimental

The behaviour of "Sigmaguard CSF", a commercial high solids content and solvent free epoxy paint produced by Sigma Coatings, has been studied. Table 1 gives the most important technical characteristics of this paint. Naval steel, St 35.8, has been employed as metallic substratum. The samples studied were rectangular test pieces of 100x150x2 mm, dry pressure blasted until a finish of type ISO-Sa 2 1/2 was obtained. The paint was then applied by airless spray to a thickness of approximately 300μm, the optimum thickness recommended in the technical data sheet of the paint.

Table 1: Main characteristics of the SIGMAGUARD CSF paint.

Description	two component solvent free amine cured epoxy coating
Colours and gloss	green-gloss
Mass density	approx. 1.3 g/cm^3
Solids content	100% by volume
VOC (supplied)	max. 17 g/l
Recommended dry film thickness	300 μm
Theoretical spreading rate	3.3 m^2/l for 300 μm
Flash point	base and hardener: above 65°C

The coating protective capacity was evaluated by monitoring the painted samples behaviour in immersion tests by means of electrochemical techniques. For this purpose, electrochemical cells were constructed, in which an area of 13 cm^2 of the painted surfaces was exposed. In some experiments, as ENM or EIS with two electrodes, two electrochemical cells were coupled and joined by a saline bridge. A solution of HCl 3M served as aggressive medium, a saturated calomel electrode was used as reference electrode and a graphite bar, as counter-electrode.

2.1 Electrochemical impedance spectroscopy

A 1287 potentiostat coupled to a 1255 frequency response analyser (FRA), both from Solartron, was used for the impedance measurements. The frequency measurement range employed for recording the impedance spectra was 10^5 to 10^2 Hz and the signal amplitude was 20 mV versus the corrosion potential. Electrical circuits were fitted with Zview software, using non-lineal method of minimums square.

Impedance measurements were recorded separately for both the coupled system and each coated sample (Cell 1 and Cell 2), so that the evolution of the coating degradation could be monitored.

2.2 Electrochemical noise measurements

The Electrochemical Noise potential and current signals were measured simultaneously by means of the classical configuration for this technique [3,4]. A Solartron 1287 potentiostat controlled by the Corrware program was employed to carry out the experiments.

In order to average the values of the parameters analysed, four consecutive Electrochemical Noise records were measured in each exposure time. 2048 points were taken in each record, at a sampling velocity of 2.15 points per second; therefore the time of each record was approximately 15 minutes.

3 Results and discussion

3.1 Electrochemical impedance spectroscopy

A change in the dielectric properties of a coating takes place when they are exposed to a corrosive media. Figure 1 represents some of the impedance spectra corresponding to tests made on painted metals immersed in acid media after being subjected to different periods of time.

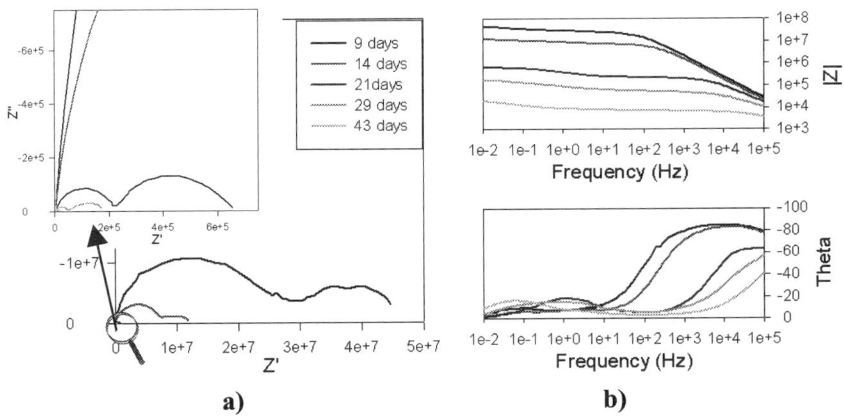

Figure 1: EIS of coated samples immersed in acid media after some days of exposure: a) Nyquist diagrams; b) Bode diagrams.

The data obtained on coated steel show a general decrease in the modulus of the impedance (Bode diagrams) along the immersion time, due to the degradation occurred in the paint (Figure 1). As we can see in the Nyquist diagrams of this figure, spectra are characterised by presenting two semicircles after the day 9. Because of this characteristic, the equivalent circuit selected for simulating the response of the system was the classical for a painted metal (Figure 2). In coated metal samples, the high frequency interval is usually related to dielectric properties of organic coatings; while low frequency gap corresponds to metal performance during the corrosion process.

In the equivalent circuit of Figure 2, each electrical element has a physical meaning [5]. R_e represents electrolyte resistance, that is, the resistance between the working electrode and the reference electrode. C_c is the capacity of the coating film and its increase, at the initial moment, is related to absorption of water or the permeability degree of the paint. R_{po} is pore resistance and is a measure of the degradation level that is suffering the paint.

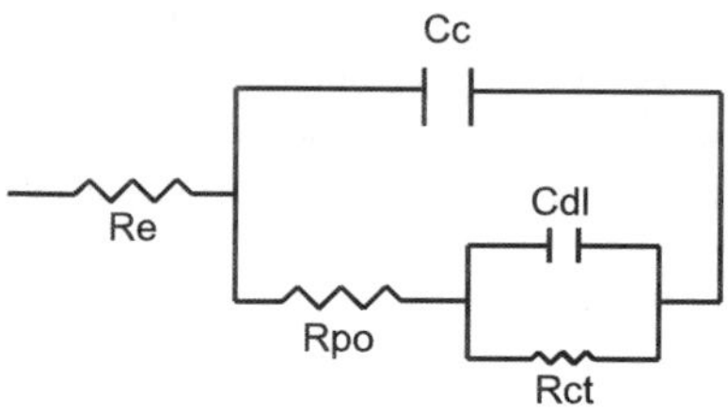

Figure 2: Equivalent circuit for painted metals.

Faradaic impedance, related to the second constant time of the spectra, represents electrochemical process taking place in the metal-coating interface. C_{dl} is the double-layer capacity and is a measure of the paint disbondment. R_{ct} symbolizes the charge transfer resistance, associated to the corrosion kinetic process. This parameter is proportionally inversed to corrosion rate and to the area where oxidation occurs. Its natural tendency is to decrease along the time of permanence in dissolution.

As a general rule, a metal-coating system that has good performance against corrosion is characterised by presenting high values of R_{po} and R_{ct} and low values of C_c and C_{dl} [6]. It does not exist a general agreement, but according to some authors [4,7,8], values of pore resistance lower than 10^7 or 10^6 Ωcm^2 reveal that coatings have a bad behaviour in relation to metal protection. Likewise, other authors [9] consider that is more important when the second semicircle appears (electrochemical process), than the values reached by the elements of the circuit.

From fitting of spectra (Fig.1) to selected electrical circuit (Fig.2), values of Table 2 are obtained. For first days, just first semicircle was still defined, so only the elements from first net (C_c and R_{po}) were fitted.

As it was expected, elements C_c and C_{dl} increase in relation to time evolution, while R_{po} and R_{ct} decrease along immersion time; tendencies related to the physical meaning explained before. According to the values of R_{po} cited and to the time when the second semicircle appears, it can be considered that our coating provides an excellent anticorrosive behaviour for the first two days of exposure to HCl. It has a questionable protection against corrosion in a period of 9 to 14 days. Meanwhile, after 14 days of immersion in HCl 3M media, the coating does not protect adequately the steel.

Table 2: Evolution of electrical elements from equivalent circuit.

Time (days)	Cc (F/cm^2)	Rpo (Ωcm^2)	Cdl (F/cm^2)	Rct (Ωcm^2)
1 hour	$4{,}17x10^{-11}$	$3{,}35x10^{9}$		
2	$6{,}86\ x10^{-11}$	$7{,}12x10^{8}$		
7	$1{,}01x10^{-10}$	$4{,}11x10^{7}$		
9	$1{,}13x10^{-10}$	$2{,}19x10^{7}$	$5{,}02x10^{-8}$	$3{,}42x10^{7}$
14	$1{,}65x10^{-10}$	$6{,}62x10^{6}$	$1{,}20x10^{-7}$	$5{,}95x10^{6}$
16	$5{,}01x10^{-10}$	$4{,}01x10^{5}$	$5{,}15x10^{-7}$	$5{,}01x10^{5}$
21	$1{,}10x10^{-9}$	$2{,}22x10^{5}$	$1{,}01x10^{-6}$	$4{,}26x10^{5}$
24	$2{,}14x10^{-9}$	$1{,}33x10^{5}$	$2{,}03x10^{-6}$	$3{,}16x10^{5}$
27	$3{,}59x10^{-9}$	$8{,}88x10^{4}$	$3{,}66x10^{-6}$	$2{,}29x10^{5}$
29	$5{,}79x10^{-9}$	$5{,}02x10^{4}$	$7{,}81x10^{-6}$	$1{,}55x10^{5}$
33	$5{,}62x10^{-9}$	$2{,}27x10^{4}$	$2{,}32x10^{-5}$	$9{,}41x10^{4}$
36	$2{,}22x10^{-9}$	$2{,}55x10^{4}$	$3{,}76x10^{-5}$	$7{,}76x10^{4}$
43	$1{,}53x10^{-8}$	$6{,}93x10^{3}$	$1{,}60x10^{-4}$	$2{,}10x10^{4}$

On the other hand, parameter $Z_{0.1Hz}$ may be used to obtain a quick evaluation of the status of the coating barrier properties. This measurement represents total impedance of the metal-coating system at frequency 0.1 Hz and some authors consider it as the optimum parameter for evaluating protection properties of coatings [2]. High values of $Z_{0.1Hz}$ are therefore associated to a high protection capacity of the coating. In general, coatings with values of $Z_{0.1Hz} > 10^8\ \Omega cm^2$ are considered as providers of an excellent anticorrosive protection; while values of $Z_{0.1Hz} < 10^6\ \Omega cm^2$ are presented by coatings with poor anticorrosive protection [10]. Figure 3 includes the evolution of parameter $Z_{0.1Hz}$ for time periods measured. We can observe that it decreases rapidly along exposure time.

According to Figure 3 and taking into consideration the criteria previously mentioned, after the day 14 the coating presents poor anticorrosive protection. As a result, we have reached to a good agreement between criteria of R_{po} element and $Z_{0,1Hz}$ parameter. This because of $Z_{0.1Hz}$ parameter, which is a good measurement for evaluating coating barrier properties applied to metals. In the one hand, if the coating is protecting adequately the metal, R_{po} element from the equivalent circuit can be approximated to $Z_{0.1Hz}$ parameter. On the other hand, when the coating does not provide protection enough to the metallic substratum, the referred parameter is associated to the corrosion process (R_{ct}). Since these resistances acquire different values, $Z_{0.1Hz}$ constitutes a good and rapid to obtain parameter of the coating degradation state.

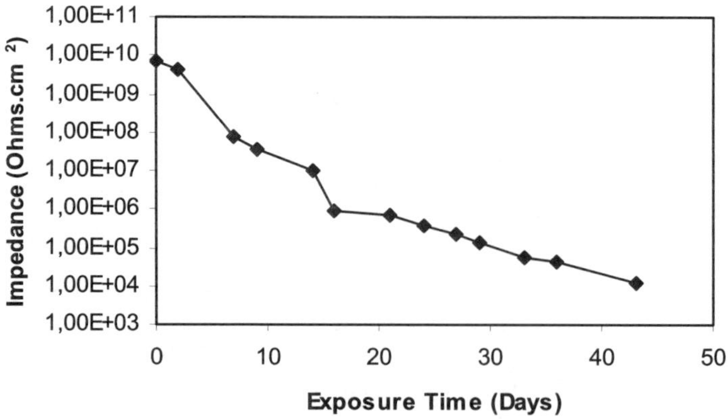

Figure 3: Evolution of $Z_{0,1Hz}$ parameter versus immersion time.

3.2 Electrochemical noise

In Figure 4 the potential and current mean values of the records have been plotted in function of the immersion time. In this figure it can be seen that during the first two days the system has low absolute potential values (about -0.3V), while after seven days the potential suddenly falls to more active values (below -0.6 V), closer to free potential of bare steel. This decrease can be related to a loss in the coating protective properties, since it is usually accepted that the more active the corrosion potential is, the more susceptible to oxidation is the metal surface under the coating [11].

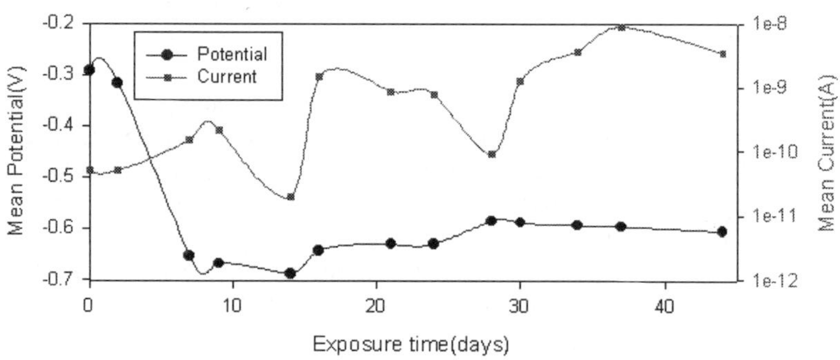

Figure 4: Potential and current mean values at different exposure days.

Figure 4 shows the influence of the current mean values (parameter called "coupled current", I_c) versus the immersion time. Although the I_c theoretical value is zero, there are some experimental deviations of this value because of different activities of the two working electrodes. In several papers it can be found that the higher activities of the systems (high corrosion rates), the higher the I_c values [12-16]. It can be appreciated from plot that I_c values increase notably in relation to the immersion time (see the logarithmic scale). This tendency reflects that the system activity increases according to the exposure time.

In order to verify that the activity of the studied system increases versus the exposure time, noise resistance (R_n) values of the time records were estimated. This parameter is one of the most commonly used for the interpretation of electrochemical noise [17,18]. In several cases, R_n is equivalent to the polarisation resistance (R_p) measured with linear polarisation [19,20], therefore it is proportionally inversed to the corrosion rate, as R_{ct} element is. R_n can be estimated dividing the standard deviation of the potential by the standard deviation of the current [21-22]:

$$R_n = \frac{\sigma_E}{\sigma_I} \tag{1}$$

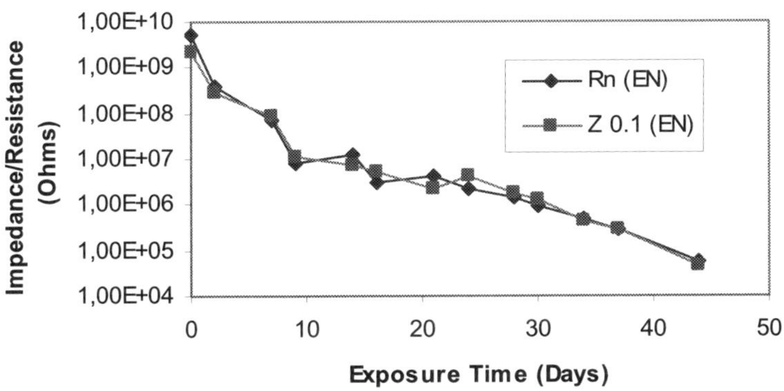

Figure 5: Comparison of Rn and $Z_{0.1Hz}$ values measured with EN.

In Figure 5 noise resistance values of the EN records have been included. In this figure the values of the impedance at 0.1 Hz, $Z_{0.1Hz}(EN)$, calculated using EN, have been added with the aim of being compared with the R_n values. $Z_{0.1Hz}(EN)$ parameter was calculated using the estimation of the potential and current PSD (equation 2). PSD were estimated by means of FFT.

$$Z_{0.1Hz}(EN) = \frac{\sqrt{PSD}_E}{\sqrt{PSD}_I}(0.1Hz) \tag{2}$$

R_n values were found to be very similar to $Z_{0.1Hz}(EN)$ values. When these parameters are analysed, the assumption derived from the analysis of the mean current (I_c) and potential values can be confirmed. That is, the longer the exposure time is, the lower the values of R_n and $Z_{0.1Hz}(EN)$ are, this means that the activity of the system is higher and the protection provided by the coating is lower.

Values of R_n equal or higher than 10^9 $\Omega.cm^2$ are typically measured for excellent coatings [23-25]. However, the line that divides good and poor behaviour is not clear. Some authors claim that a coating presenting a R_n value below 10^6 $\Omega.cm^2$ is not protective [25], while others think that the line should be drawn at 10^7 $\Omega.cm^2$ [26]. According to these considerations, it can be stated that the coating studied showed excellent anticorrosive behaviour before the second day of immersion, good behaviour in a period of 2 to 7 days and poor behaviour after 14 days of exposure in HCl.

3.3 Correlation between EIS and EN

Values of parameters $Z_{0.1Hz}$, calculated using both EIS and EN, and R_n values have been included in Figure 6 to be compared. When these parameters are analysed, the expected tendency is confirmed. Therefore, the higher the exposure time is, the lower the values of all these parameters are. This evolution is related to the higher activity in the system and the lower protection provided by the coating mentioned before. In addition, it can be observed that an excellent agreement exists between the parameters calculated with EIS and EN. In this figure it can be appreciated that $Z_{0.1Hz}$ parameter of the double cell (calculated by EIS) is very similar to $Z_{0.1Hz}$ values of the cell with lower impedance (Cell 1 when the exposure time is lower than 16 days and Cell 2 when the time of immersion is higher than 16 days). Therefore, the double cell reflects the electrochemical behaviour of the single cell that behaves worse.

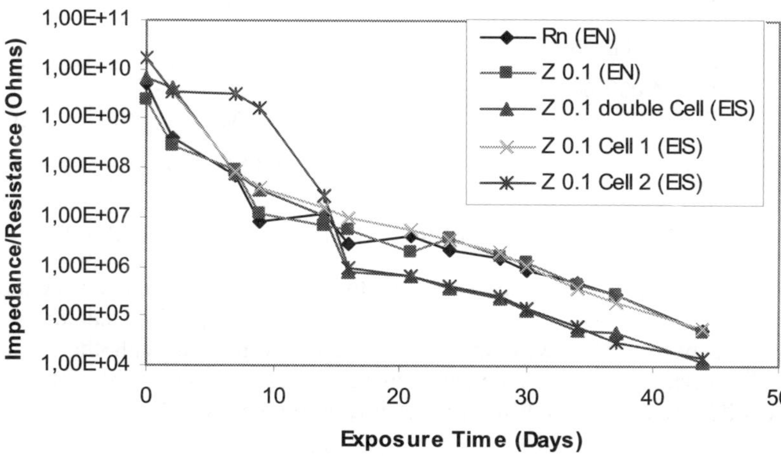

Figure 6: Comparison of Rn and $Z_{0.1}$ values measured with EIS and EN.

On the other hand, since EN technique needs the double cell to carry out the experiments, this technique measures impedance values similar to the ones calculated in the double cell with EIS. According to the considerations mentioned before, EN measures the electrochemical properties of the electrode with lower impedance, fact that ensures the EN technique as very sensitive to measure decreases of impedance in coated electrodes.

4 Conclusions

In the present work a two-component solvent free amine cured epoxy coating has been used to protect steel. Carbon steel samples were coated with this system and later on immersed in HCl, a very aggressive solution employed to accelerate degradation. Electrochemical Impedance Spectroscopy was used to discriminately analyse the electrochemical response of the metal-paint system. Electrochemical Noise Measurements were also carried out and interpreted. Excellent correlation among EIS and EN data was found.

According to the criteria previously exposed, the coating studied showed excellent anticorrosive behaviour before the second day of immersion, good behaviour in the period of 2 to 7 days, and poor behaviour after 14 days of exposure.

References

[1] Stoye, D. & Freitag, W. *Paints, Coatings and Solvents*, Wiley-VCH: Weinheim (Germany), pp. 105-109, 1998.

[2] Weiss, K.D., Paint and coatings: a mature industry in transition. *Progress in Polymer Science*, 22, pp. 203-245, 1997.

[3] Aballe, A., Bethencourt, M., Botana, F.J., Marcos, M. Using wavelet transform in the analysis of electrochemical noise data. *E. Acta*, 44, pp. 4805-4816, 1999.

[4] Aballe, A., Bethencourt, M., Botana, F.J., Marcos, M., Sanchez-Amaya, J.M. Use of wavelet to study electrochemical noise transients. *E. Acta*, 46, pp. 2353-2361, 2001.

[5] Amirudin, A. & Thierry, D., Application of electrochemical impedance spectroscopy to study the degradation of polymer-coated metals. *Progress in Organic Coatings*, 26, pp. 1-28, 1995.

[6] Tang, N., van Ooij, W.J. & Górecki, G., Comparative EIS study of pre-treatment performance in coated metals. *Progress in Organic Coatings*, 30, pp. 255-263, 1997.

[7] McIntyre, J.M., Electrochemical impedance spectroscopy; a tool for organic coatings optimizations. *Progress in Organic Coatings*, 30, pp. 255-263, 1997.

[8] Deflorian, F., Rossi, S., Kamarchik, P., Fedrizzi, L. & Bonora, P.L., Degradation mechanism of electrodeposited coatings in alkaline solution. *Progress in Organic Coatings*, 27, pp. 201-207, 1996.

[9] del Amo, B., Romagnoli, R., Deyá, C. & González, J.A., High performance water-based paints with non-toxic anticorrosive pigments. *Progress in Organic Coatings*, 45, pp. 389-397, 2002.

[10] O'Donaghue, M., Garret, R., Vijay, D., Roberts, P. & T. Aben, Electrochemical Impedance Spectroscopy: Testing Coatings for Rapid Immersion Service. *Materials Performance*, Sept. pp. 36-40, 2003.

[11] Santagata, D., Seré, P.R., Elsner, C. & Di Sarli, A., Evaluation of the surface treatment effect on the corrosion performance of paint coated carbon steel. *Progress in Organic Coatings*, 33(1), pp. 44-54, 1998.

[12] Schmitt, G., Moeller, K. y Plagemann, P. A new service oriented method for evaluation of electrochemical noise data for online monitoring of crevice corrosion. *Proceedings of Corrosion/2004*, Paper 04454, NACE. pp. 1-10. New Orleans (Louisiana), U.S.A., 2004.

[13] Aballe, A., Bethencourt, M., Botana, F.J., Marcos, M. y Sánchez-Amaya, J.M. ENM study of corrosive systems showing high drift. *Proceedings of 15th International Corrosion Congress.* Paper Nº 566. International Corrosion Council pp. 1-8. Granada, 2002.

[14] Mansfeld, F. and Xiao, H. Electrochemical Noise and Impedance Analysis of Iron in Chloride Media. Electrochemical Noise Measurement for Corrosion Applications, Ed. ASTM., West Conshohocken, U.S.A., pp. 59-78,1996.

[15] Edgemon, G.L., Danielson, M.J., Bell, G.E.G. Detection of stress corrosion cracking and general corrosion of mild steel in simulated defense nuclear waste solutions using electrochemical noise analysis. *Journal of Nuclear Materials*, 245, pp. 201-209, 1997.

[16] Greisiger, H. and Schauer, T. On the interpretation of the electrochemical noise data for coatings. Progress in Organic Coatings, 39, pp. 31-36, 2000.

[17] Cottis, R.A. Interpretation of Electrochemical Noise Data. *Corrosion*, 57, pp 265-285, 2001.

[18] Huet, F., Bertocci, U., Gabrielli, C. and Keddam, M. Noise Measurement in corrosion. *Proceedings of Corrosion/97*, NACE, pp. 11-30 New Orleans, U.S.A, 1997.

[19] Bautista, A. and Huet, F. Noise Resistance Applied to Corrosion Measurements: IV. Asymmetric coated electrodes. *J. Electrochem. Soc.* 146 pp. 1730-1736, 1999.

[20] Bautista, A., Bertocci, U. and Huet, F. Noise Resistance Applied to Corrosion Measurements: V. Influence of the electrode asymmetry. *J. Electrochem. Soc.* 148 pp. B412-B418, 2001.

[21] Bertocci, U. and Huet, F. Noise analysis applied to electrochemical systems. *Corrosion* 51, pp. 131-144, 1995.

[22] Gabrielli, C., Huet, F. and Keddam, M. Investigation of electrochemical processes by an electrochemical noise analysis. Theoretical and experimental aspects in potentiostatic regime. *Electrochim. Acta* 31, pp. 1025-1039, 1986.

[23] Le Thu, Q., Bierwagen, G. P. and Touzain, S. EIS and ENM measurements for three different organic coatings on aluminium. Progress in Organic Coatings, 42, pp. 179-187, 2001.

[24] Chen, C.T. and Skerry, B.S. Assessing the corrosion resistance of painted steel by AC impedance and electrochemical noise techniques. *Corrosion,* 47, pp. 598–611, 1991.

[25] Bierwagen, G.P., He, L., Li, J., Ellingson, L., Tallman, D.E. Studies of a new accelerated evaluation method for coating corrosion resistance-thermal cycling testing. *Progress in organic coatings*, 39, pp.67-78, 2000.

[26] Mills, D. J. and Mabbutt, S. Investigation of defects in organic anti-corrosive coatings using electrochemical noise measurement. Progress in Organic Coatings, 39, pp. 41-48, 2000.

Simulation of the formation mechanism of a viscous layer for the electropolishing process

S.-J. Lee, J.-J. Lai & Y.-T. Lin

Department of Mechanical Engineering, Yuan Ze University, Taiwan

Abstract

Electropolishing (EP) is a surface finishing treatment method. The surface quality indices such as surface roughness, cleanness and corrosion resistance could be improved through this process. Because of these benefits, the electropolishing process is applied to fabricate high cleanness apparatus for semiconductor and optical pharmaceutical industries.

Electropolishing is an electrochemical reaction process. The uniform and stable passive film is the key mechanism in enhancing the surface quality. However, the forming of passive film is influenced by the current distribution, fluid field and diffusion phenomena. Therefore, there are many defects such as pits, flow marks and scrape which could be found after the electropolishing process. These defects destroy the uniformity of the surface quality and induce local corrosion.

The goal of this study is aimed at simulating the formation of a viscous layer and establishing numerical a model to predict its thickness and uniformity. In this study, the commercial software of FEMLAB was employed to establish a multi-physics model which contained the effects of fluidity and diffusion. The mechanism of passive film formation, the influence of fluid velocity and the thickness variation of the viscous layer were investigated through numerical simulation.

Keywords: electropolishing, passive film, numerical simulation, multi-physics.

1 Introduction

The electropolishing is a precision surface finishing treatment method. It could enhance the cleanliness and smoothness of a specimen. The thin passive film formed on the surface will also improve its corrosion resistance [1-3]. The process is composed of an anode (specimen), a cathode and acidic electrolyte. During the process, hydrogen will be generated on the cathode while oxygen on the anode. The metallic ions react with the electrolyte to form diffusion layer, also called viscous layer, on the surface. This diffusion layer is the key factor that controls the leveling, polishing and passivity mechanisms of the electropolishing process.

Many researchers define the diffusion process as the controlling mechanism of the EP process [4-6]. However, there are other effects such as convection, temperature distribution, gas bubbles and electron migration taking place at the same time which makes the whole process multi-physics phenomena. Hence, the control of the initiation and stability of the diffusion process becomes very complicated. Eventually, they affect process parameters of electrode gap, polishing time, applied potential and electrode design.

Therefore, the diffusion layer formed under diffusion control is fragile and unstable. It may be easily affected by the process and has become the key factor in achieving process control and product quality. The method of rotating disc electrode (RDE) was proposed and was reported to have beneficial effects on the EP process [7-8]. Mechanical agitation of the electrolyte seemed to also have the same effects [9-10]. Many research results pointed out refreshing of the electrolyte by fluid control can improve EP polishing effects and process stability.

However, due to the lack of understanding on the formation mechanism of the diffusion layer, various researches chose vastly different process parameters for their experiments [11-13]. In order to gain deeper insight, this paper employed FEMLAB, a commercial code, to simulate the effects of diffusion and convection on the diffusion layer.

2 The problem statement

During the EP process, the ionic concentration on the electrode surface is varying. For example, the metallic ions will dissolve from the anode and react with electrolyte. This will increase the concentration of metallic ions on anode surface. The concentration gradient results in the formation of diffusion layer which, in turn, creates the plateau region of the I-V curve.

In this research, numerical simulations were performed to study the effects of the electrolyte flow to the diffusion layer. The schematic plot of the electropolishing process was shown in Figure 1(a). Assumptions of incompressible fluid, laminar flow and constant temperature were made in order to simplify the numerical model.

In the electrochemical reactions, the transport phenomenon was composed of the ionic migration, diffusion, convection and electron flow. If the concentration of electroactive species is low, the interaction between molecules could be ignored. Hence, the mass transfer in the electrolyte occurs under several physical phenomena; ionic migration under the influence of electric field, diffusion due to concentration grading, and convection. It is shown as follows:

$$Ni = -z_i u_i F C_i \nabla \phi - D_i \nabla C_i + C_i v \qquad (1)$$

On the right hand side, they represent the effects of migration, diffusion and convection. Where z_i, u_i are the number of carrying electron and its coefficient of mobility for specie i, $\nabla \phi$ is the grading of electric field, C is the concentration and F is the Faraday's constant.

During the formation of diffusion layer, the polarization phenomena also take place simultaneously which will decrease the effect of migration. However, the fluid dynamics has much greater effects on the formation of diffusion layer and its stability. Therefore, the effects of electron migration were ignored in this study. The effect of gas bubbles makes the fluid dynamics a two-phase flow problem which increases the complexity on numerical simulation considerably. Therefore, it was also ignored. The numerical model had been simplified to the study of diffusion and convection under various fluid field velocities.

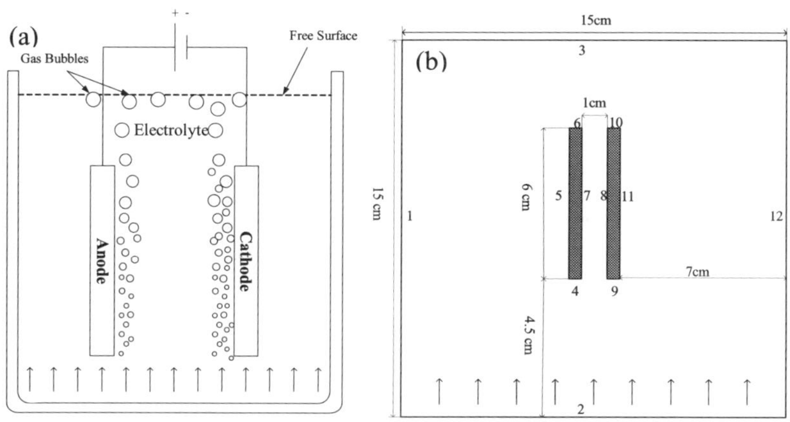

Figure 1: (a) Schematic picture of the EP process. (b) The specifications of the model.

2.1 The model specification

This study mainly concerns the phenomena of diffusion layer in between the electrode gap. Therefore, problem definition was focused around the electrode area. The geometry and specification of the 2D model was presented in Figure 1(b). The model specifications remained the same throughout this study. The entire flow fluid area is 15 cm by 15 cm. The area of both anode and cathode was 0.6 cm^2. A jet flow pattern from the bottom of the tank, boundary no. 2, with uniform speed was assumed. The boundaries were assigned with numerical number, as showed in Figure 1(b), for boundary specifications.

2.2 The diffusion and convection model

The Fick's law is the governing equation in diffusion and convection model, shown in Equation 2. Since steady state is assumed, hence, the rate for chemical reaction is 0, shown in Equation 3.

$$\frac{\partial C}{\partial t} + \nabla(-D\nabla C + Cu) = R \tag{2}$$

$$\frac{\partial C}{\partial t} + \nabla(-D\nabla C + Cu) = 0 \tag{3}$$

where C: concentration of diffusive species, D: diffusion coefficient, u: velocity of fluid, R: rate of chemical reaction.

The diffusion mechanism of the EP process is defined as ordinary diffusion liquid. Because it is very hard to measure the diffusion coefficient, the diffusion coefficient was estimated from the Storkes-Einstein equation, as presented in Equation 4. Although the precision of the Storkes-Einstein equation is not good, it is our best available tool.

$$D = \frac{\kappa T}{6\pi\mu R} \tag{4}$$

where D: diffusion coefficient, κ :Boltzmann's constant, μ :Viscosity of electrolyte, R: Solute radius.

2.3 The fluid field model

Because turbulence will affect the uniformity of the reaction rate and result in lack of process control, it should be avoided in the flow of electrolyte. Therefore, the conventional Navier-Stokes Equation is adopted as the governing equation for the fluid field model. Equation 5 represents laminar fluid field in the tank of electrolyte.

$$\begin{aligned} &\rho\frac{\partial u}{\partial t} - \eta\nabla^2 u + \rho(u\cdot\nabla)u + \nabla p = F \\ &\nabla\cdot u = 0 \end{aligned} \tag{5}$$

where ρ : density, η : dynamic viscosity, u: velocity field, p: pressure, F: volume force.

2.4 The boundary conditions

In the laminar flow model, no slip condition, $u = 0$, is assumed for all boundaries except the inlet and outlet. The inlet, boundary no. 2, is assigned as inflow with constant flow speed in Y direction. In the diffusion and convection model, the reactant formed on the anode will diffuse toward the cathode. Hence, the concentration on the anode surface, boundary no. 7, was assumed to be one while it is zero at the cathode, boundary no. 8. The other boundaries, boundaries no. 1, 2 and 4-12, were not involved in the electrochemical reaction. Their flux values were assigned to be zero, $-D\nabla c + cu = 0$. Because the effect of convection is much stronger than diffusion at the outlet, boundary no. 3, there are outflow of flux from the outlet. Therefore, it is defined as convective flow, $-D\nabla c = 0$. The settings of boundary conditions were listed in Table 1.

Table 1: The settings of Boundary conditions.

Boundary settings of the flow model			
Boundary no.	1,4-12	2	3
Description	No slip	Inflow	Outflow
Setting	u=0	$u=(u_0,v_0)$	P=0
Boundary settings of the diffusion and convection model			
Boundary no.	1,2,4-12	3	7
Description	Flux	Convective flux	concentration
Setting	Flux=0		1

2.5 Computational parameters

Three levels of diffusion coefficient were selected to study the effects of its interaction with fluid field on the thickness and shape of the diffusion layer. They were 10^{-7}, 10^{-8} and 10^{-9}cm/s, respectively. In order to ensure laminar flow, the velocity of fluid should be slow. The initial speeds at the inlet were assigned as 1cm/s, 0.5cm/s and 0.1cm/s. In order to have the same bases for comparison, the simulation duration was set at 300 seconds for all analyses.

3 Results and discussions

First, the numerical simulations with no fluid field were performed to be taken as the references. Three levels of diffusion coefficient were considered in the simulation. When the level of concentration is higher than 90%, it was considered as the viscous layer. Because the viscous layer is the most important mechanism in controlling the EP process, the variations in the thickness and shape of the viscous layer were the focus in this study. Figures 3, 4 and 5 represented the growth of diffusion layer over time for diffusion coefficients 10^{-7}, 10^{-8} and 10^{-9}cm/s, respectively.

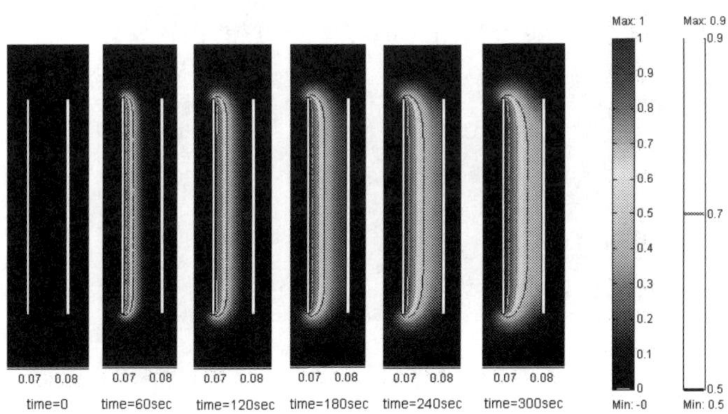

Figure 2: Profile of the diffusion layer, $D=10^{-7}$cm/s.

In Figure 2 where the value of diffusion coefficient was 10^{-7}cm/s, the diffusion process is still in transient state after 300 seconds. A steady state is not reached. The diffusion layer is still changing with time. The results may reflect the fact that the impendence variation between the electrode gaps is too large for the EP process. Therefore, the process is hard to control and the uniformity may not be good.

When the diffusion coefficient is 10^{-8}cm/s, shown in Figure 3, the diffusion speed is slower. It reached steady state in 180 seconds. It may indicate that the impedance in between the electrode gap is more stable. The thickness of the diffusion layer is thin and dense. For the leveling mechanism of the EP process, a thin and dense diffusion layer provides better selective dissolution and, thus, better polishing results and uniformity.

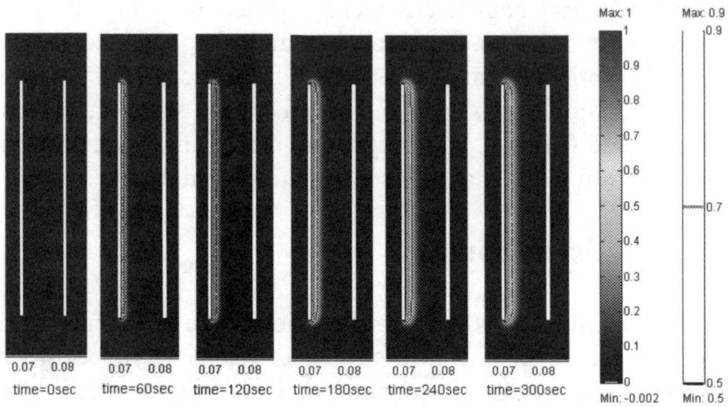

Figure 3: Profile of the diffusion layer, D=10^{-8}cm/s.

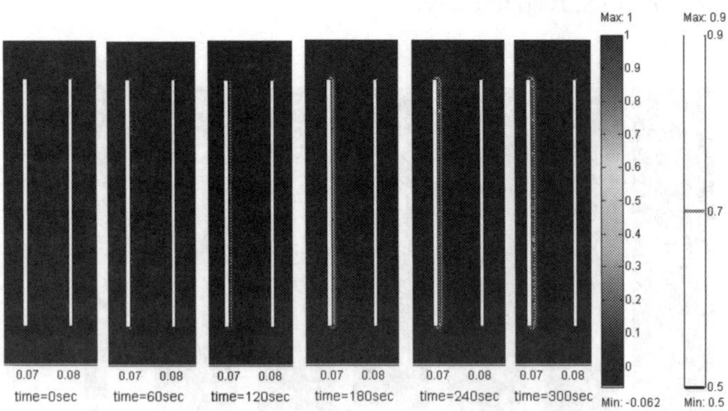

Figure 4: Profile of the diffusion layer, D=10^{-9}cm/s.

When the diffusion coefficient is as low as 10^{-9}cm/s, it is very hard to diffuse, as demonstrated in Figure 4. It also reached steady state in 180 seconds. However, the thickness of the diffusion layer was only a few tens of micrometers. Ideally, it may result in very smooth surface. However, it is very hard to maintain a uniform diffusion layer of such thin thickness. Therefore, it is estimated that the diffusion coefficient of the EP process should be between 10^{-7}cm/s and 10^{-8}cm/s.

The following simulation analyses of fluid field on the EP process will be performed on these two values of diffusion coefficient.

3.1 Results of fluid field simulation with diffusion coefficient D=10^{-7}cm/s

Figure 5 showed the simulation results of flow speed equals to 1 cm/s. The diffusion layer was stabled down after 60 seconds. Comparing Figure 5 to Figure 2, the stability of the impedance in between the electrode gap had shown significant improvement. However, there were greater variation at both the inlet and the outlet due to stronger fluid dynamics effect. From the contour line of 90% concentration level of Figure 5, it was found that the thickness of the viscous layer had grown from 0.05mm at the inlet to 0.1 mm at 1mm distance along the Y-axis toward the outlet. It had grown further to 0.2mm at Y=3.8mm, to 0.3mm at Y=13.2mm, to 0.5mm at Y=17mm and finally to 2mm at the outlet. The difference in the thickness of the viscous layer was ten folds. It will result in differences in the metallic dissolution rate at both the inlet and the outlet and, thus, the lack of uniformity of the overall EP polishing results.

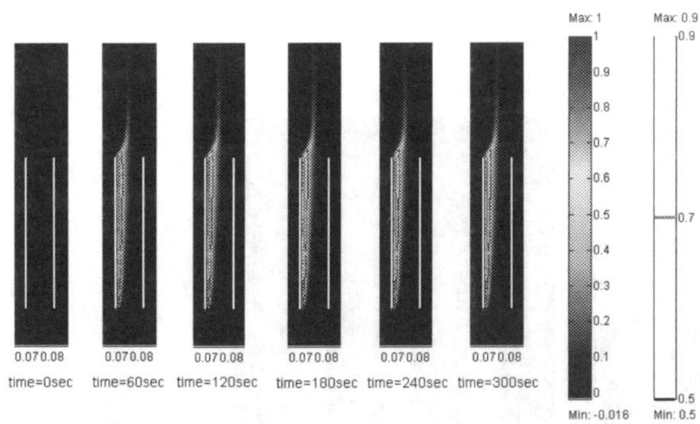

Figure 5: Profile of the diffusion layer, D=10^{-7}cm/s, V=1cm/s.

Figure 6 showed the simulation results of flow speed equals to 0.5cm/s. The diffusion layer had reached stable state after 120 seconds. However, it had the same problems of significant variation in the thickness of the viscous layer between the inlet and the outlet. The difference in between the thinnest and the thickest was also ten folds.

The diffusion layer had reached stable state after 180 seconds for the simulation results with flow speed equates to 0.1cm/s, as shown in Figure 7. The distribution of viscous layer was from 0.1mm to 0.9mm in thickness. The difference was only nine folds, the least in these three sets of simulation analyses. The thickness and shape of the viscous layer were better in comparison with Figures 6 and 7.

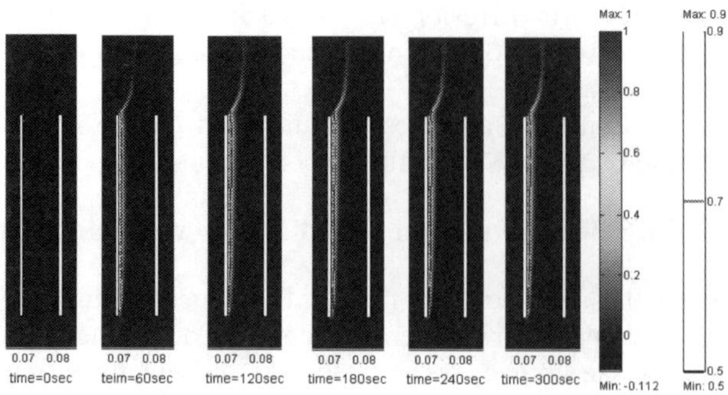

Figure 6: Profile of the diffusion layer, D=10^{-7}cm/s, V=0.5cm/s.

From the above simulation results, it could be concluded that, when the value of the diffusion coefficient is 10^{-7}cm/s, the fluid field could significantly reduce transient time. Shorter transient time meant less variation in the impedance in between electrode gap and, thus, better process uniformity. However, higher flow speed will result in significant differences in the thickness of the viscous layer. A ten folds difference was computed for the flow speed equals to 1cm/s. Therefore, slower flow speed was preferred.

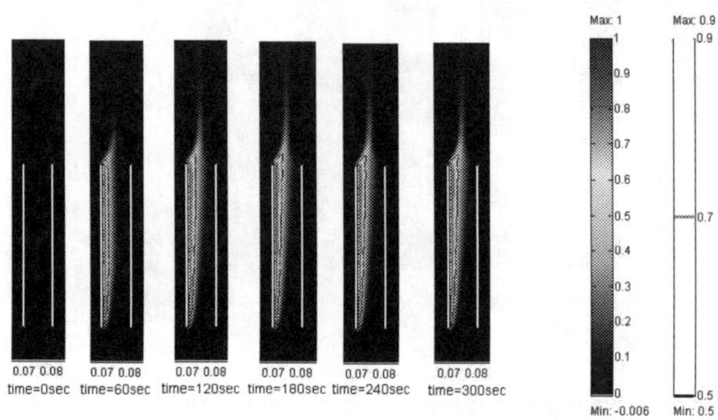

Figure 7: Profile of the diffusion layer, D=10^{-7}cm/s, V=0.1cm/s.

3.2 Result of fluid field simulation with diffusion coefficient D=10^{-8}cm/s

Figure 8 showed the simulation result of flow speed equals to 1cm/s. The results indicated that stability of the diffusion layer could be reached in 60 seconds, a significant reduction from 120 seconds with zero flow speed. The thickness of the viscous layer was 0.05 mm at the inlet and was 0.1 mm thick at Y=5.4 mm. It reached 0.2 mm thick at Y=42.4 mm and maintain stable afterward. The difference between the inlet and the outlet was only four folds. The uniformity was very much improved from the results with diffusion coefficient D=10^{-7}cm/s where the difference is ten folds.

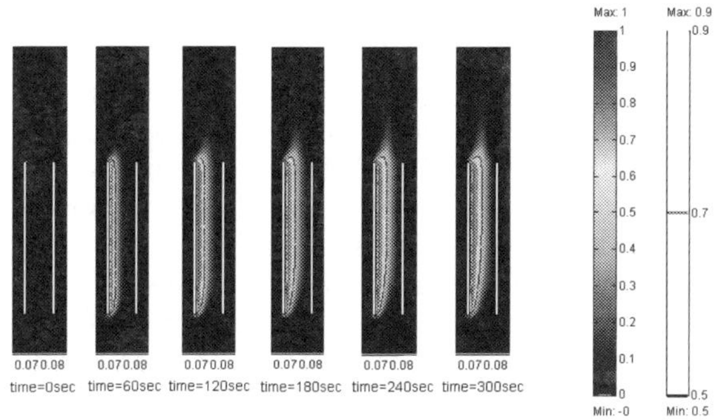

Figure 8: Profile of the diffusion layer, D=10^{-8}cm/s, V=1cm/s.

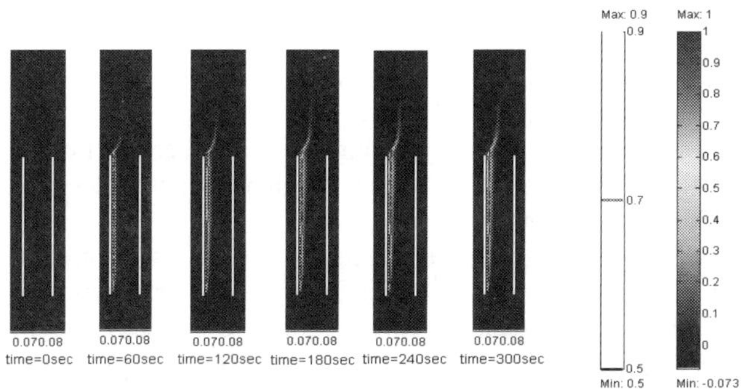

Figure 9: Profile of the diffusion layer, D=10^{-8}cm/s, V=0.5cm/s.

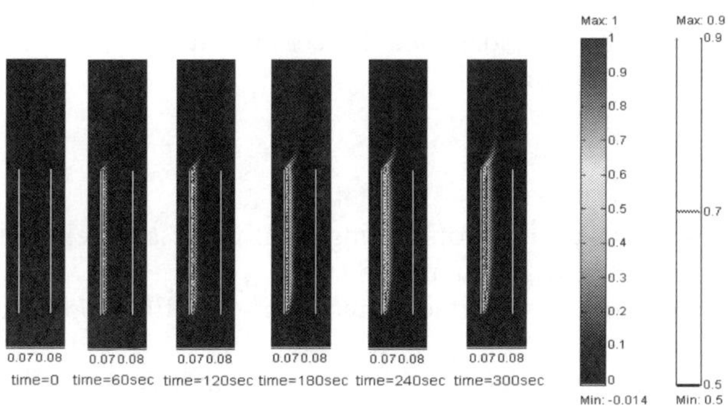

Figure 10: Profile of the diffusion layer, D=10^{-8}cm/s, V=0.1cm/s.

For flow speed equals to 0.5cm/s, the diffusion layer will reach stable state after 120 seconds, as shown in Figure 9. The distribution of the viscous layer thickness could be divided into four sections. From the 0.05 mm thick at the inlet, to 0.1mm at Y=4 mm, to 0.2 mm at Y=12 mm and maintain stable afterward. The shape and thickness of the viscous layer was even more uniform than in Figure 8.

For flow speed equals to 0.1 m/s, the diffusion layer will reach stable state after 180 seconds, as shown in Figure 10. The diffusion layer became thicker. The viscous layer reached 0.3 mm thickness after Y=2.2mm and maintain through the entire anode length. The difference in thickness between the inlet and the outlet was six folds. The shape and thickness distribution of the viscous layer was the best for all analyses results.

In summary, the results of diffusion coefficient, $D=10^{-8}$cm/s, can reached the stable state about the same time as those of diffusion coefficient $D=10^{-7}$cm/s. However, the thickness and its distribution of the viscous layer were much better. The uniformity and stability of the viscous layer improve with the decrease in flow speed. However, the thickness at the inlet was still about 0.05mm which may create minor problem in uniformity.

4 Conclusions

1. When there is no fluid field, the diffusion takes very long time which may affect the uniformity of the EP process.
2. The electrolyte should have high coefficient of viscosity in order to increase the diffusion coefficient of the solute. Higher diffusion coefficient improves uniformity of the viscous layer.
3. The fluid field has significant effects on the diffusion mechanism. Proper flow speed may effectively reduce time of transient state, improve uniformity of the viscous layer and enhance the overall polishing effects.
4. It is possible to increase process control and polishing precision by adjusting the fluid field to affect the distribution and the thickness of the diffusion layer.
5. Through the numerical simulation model, the diffusion mechanism of the EP process could be better understood. Process parameters could be analyzed to improve process control.

References

[1] Charles L. Faust, "Electropolishing — Stainless Steel Part II," Metal Finishing, pp. 53-56, February 1983.

[2] Kenneth B. Hensel, "Surface Treatments — Electropolishing," Metal Finishing, 1989.

[3] Charles L. Faust, "Electropolishing — Stainless Steel Part I," Metal Finishing, pp. 21-25, July 1982.

[4] M. Matloze, "Modeling of Impedance Mechanisms in Electropolishing," Electrochemica Acta, Vol. 40, No. 4, pp. 393-401 1995.

[5] Simon L. Marshall and Stephen K. Wolf, "Analysis of Terminal Effects in Rectangular Electrochemical cells," Electrochemica Acta, Vol. 43, pp. 405-415, 1998.

[6] Y. Wu and Z. Wang, "The Theoretical Behaviors of Tubular Electrodes: From Semi-Infinite Diffusion to Bulk Electrolyte," Electrochemica Acta, Vol. 44, pp. 2281-2286, 1999.

[7] O. Piotrowski, C. Madore and D. Landolt, "Electropolishing of Tantalum in Sulfuric Acid-Methanol Electrolytes," Electrochemica Acta, Vol. 44, pp. 3389-3399, 1999.

[8] Tadeusz Hryniewicz, "Concept of microsmoothing in the electropolishing process," Surface and Technology, pp. 75 –80, 1983.

[9] G. H. Sedahmed, M. S. Abdo and M. A. Kamal, "A Mass Transfer Study of the Electropolishing of Metals in Mechanically Agitated Vessels," Int. Comm. Heat Mass Transfer, Vol. 28, No. 2, pp. 257-265, 2001.

[10] J. E. Labarga, J. M. Bastidas and S. Feliu, "A Contribution to the Study on Electropolishing of Mild Steel and Aluminium Using Alternating Current," Electrochimica Acta, Vol. 36, No. 1, pp. 93-95, 1991.

[11] M. Matlosz and D. Landolt, "Shape changes in Electrochemical Polishing: The Effect of Temperature on the Anodic Leveling of Fe-24Cr," J. Electrochem. Soc., Vol. 136, No.4, pp. 919-929, 1989.

[12] Ray Dargis, "Chemical Polishing and Electropolishing," Products Finishing, 1989.

[13] S. J. Lee and J. J. Lai, "The Effects of Electropolishing (EP) Process Parameters on corrosion Resistance of 316L Stainless Steel," Journal of Materials Processing Technology, Vol. 140, 2003.

Section 5
Modelling stress corrosion cracking and corrosion fatigue

The roles of microstructure and mechanics in intergranular stress corrosion cracking

A. P. Jivkov, N. P. C. Stevens & T. J. Marrow
School of Materials, The University of Manchester, UK

Abstract

Previous work on the prediction of intergranular stress corrosion cracking resistance in grain boundary engineered microstructures has used two dimensional percolation models, in which the grain boundaries are assumed to be either resistant or susceptible to cracking, depending on the grain boundary character. One limitation of such models is that they do not necessarily account for the mechanical crack driving force. Further, they cannot capture experimentally observed phenomena such as the formation of isolated ductile bridging ligaments by resistant boundaries. These arise due to the three-dimensional character of crack propagation. A new mechanical crack propagation model is presented which, via finite element solutions, addresses these limitations. The model is based on a regular discrete representation of the material's microstructure and is applicable to both 2D and 3D behaviour. Results are reported for 2D-hexagonal microstructures, and are compared with percolation models. The results demonstrate the influence of stress on the crack path, as well as the influence of the rupture strain of susceptible boundaries on the crack behaviour. Further, the effect of crack bridging, which arises from ductile resistant boundaries is studied.
Keywords: intergranular stress corrosion; microstructure; mechanics; crack bridging; finite elements; Monte Carlo simulations.

1 Introduction

It has been increasingly realised during the last two decades, that the grain size is not the only meso-scale factor influencing the strength of polycrystalline materials. Two other factors have been brought to wider attention. The first is the grain boundary character distribution (GBCD) which describes the fractions of boundaries with different energies, with "special" or low-energy boundaries having higher resistance to intergranular deterioration mechanisms and "random" or high-energy boundaries having lower resistance. The second factor is the topological connectivity of "random" boundaries. In connection to these findings, the concept of grain boundary engineering has been introduced by Watanabe [1]. The primary purpose of grain boundary engineering has been to improve the bulk mechanical properties of polycrystalline materials by increasing the number of "special" boundaries. Together with applications to enhance general fracture toughness [2-4], it has been shown that the "special" grain boundaries can be much less susceptible to intergranular corrosion and stress corrosion cracking (SCC) [5-7]. Since the introduction of grain boundary engineering, characterisation of grain boundary networks has been an important topic in grain boundary engineering research [8-11]. Predictive stress corrosion cracking models aim to determine the probability of crack arrest and distribution of arrested crack lengths for given grain boundary network characteristics. Previously proposed models [12-16] used a percolation-type process to determine the probable extent of crack growth. These are binary models, i.e. a grain boundary is assumed to be either entirely resistant or entirely susceptible to SCC. In the early models [12, 13], the probability of crack advance at a junction is based on the GBCD and orientation of the grain boundary with respect to the applied stress. Later models [14-16], additionally account for the network connectivity via the distribution of triple junctions. All these models, however, cannot describe the effects of the applied stress magnitude or stress redistribution during crack evolution. Hence, they stand close to a pure geometrical percolation model, giving the critical share of "random" boundaries above which a continuous crack path across a given microstructure is always possible. For the most widely used two dimensional hexagonal cell structure [12,13,15], the critical share of resistant boundaries (i.e. the percolation threshold) is found to be around 0.65, while for the space-filling tetrakaidecahedral structure in three dimensions [1,5], the threshold is around 0.23 (see e.g. [14]). Previous 2D models also cannot account for the experimentally observed crack bridging behaviour. Bridging is created by the yielding of ductile ligaments, formed by resistant boundaries left behind the advancing crack front in the real 3D geometry. A recently proposed analytical model [17], has attempted to take into account the effects of crack bridging on the local crack tip stress intensity factor, thus mimicking 3D crack behaviour.

The objective of the present work is to combine the structural development of a percolation model with actual finite element calculations of the stress after each change in geometry due to crack advance. This allows for a more accurate simulation of crack evolution, as the probability of crack advance depends upon the actual mechanical conditions at the crack tip. This is affected by the crack propagation history, i.e. the effects of crack branching, crack bridging and redistribution of the initial stress state. The work utilises a regular 2D hexagonal cell structure. Grain boundaries are assumed to belong to either of the two classes – susceptible or resistant to stress corrosion. All boundaries belonging to one class are assigned identical mechanical properties. Susceptible boundaries are assumed to fail at crack opening displacements of the order of several nanometers. This represents a local crack propagation criterion requiring a small level of crack tip strain, suggested by experimental observations of intergranular stress corrosion [18]. Resistant boundaries are assumed to deform in a ductile manner, as demonstrated by in-situ high resolution tomographic and fractographic observations of intergranular stress corrosion cracking [19]. The particular physical mechanism of cracking and hence the time-scale of crack propagation process are not involved in this work.

2 Model description

A possible approach to modelling an assembly of hexagonal cells (grains) in a finite element environment could be based on continuum mechanics. Each cell could be tessellated into continuum elements, e.g. triangles, and connected to the neighbouring cells with interface elements representing the boundary between the two grains. Such a direct strategy would require significant computational resources for the simulations when the size of the assembly increases. A simpler structural model of the grain network is proposed and used in this work. In this model, each cell is represented by a geometrical point and its connections to the neighbouring cells are represented by linear structural members. The scheme is illustrated in Fig. 1, where a portion of a plane hexagonal mesh and its corresponding structure are shown. The finite element model consists of nodes in the centres of the grains, and plane beam elements as structural members.

Let D denote the diameter of a hexagonal cell. With respect to a fixed coordinate system (X_1, X_2), the assembly of grains studied in this work fills the rectangular region $\{-50D \leq X_1 \leq 50D,\ 0 \leq X_2 \leq 50D\}$. This region contains 7740 grains, which form 22850 internal grain boundaries, modelled by 7740 nodes and 22850 beam elements in the finite element model.

An initial crack, extending along three-grain boundaries, is introduced from the surface point at the origin of the two axes in the coordinate system and running in towards the centre of the element assembly. This is schematically shown with a dashed line on the right of Fig. 1, where the length of the pre-crack is also given as 1.25D. The geometry of the crack and the boundary conditions define an initial stress intensity factor approximately as

$$K_I \approx 1.1215\,\sigma_\infty \sqrt{\pi\,(1.25D)} \qquad (1)$$

where σ_∞ is the remotely applied stress and the stress pre-factor accounts for the edge crack geometry (see e.g. [20]).

An important question in the proposed structural model of a polycrystalline solid is the choice of geometrical and mechanical properties of structural members, so that the model represents properly the assembly behaviour. In this work, the cross section of the beam elements is chosen in such a way that their elastic properties (Young's modulus, E, and Poisson's ratio, ν) coincide with the bulk metal properties, namely E = 200 GPa and ν = 0.3, while at the same time the assembly overall deformation under considered boundary conditions is the same as the deformation of a continuous solid of the same geometry.

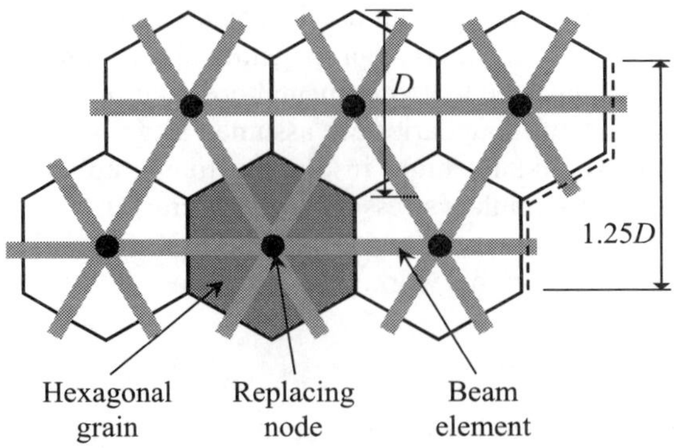

Figure 1: Illustration of the discrete model of hexagonal microstructure.

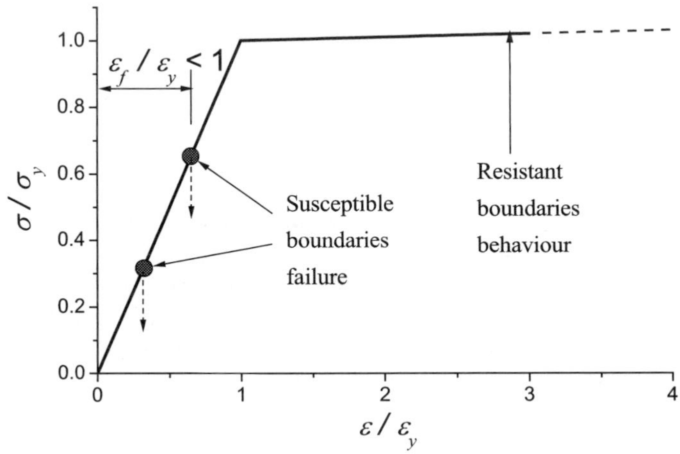

Figure 2: Schematic for constitutive behaviour of boundaries.

Two groups of grain boundaries in an assembly are considered: those susceptible and those resistant to corrosion. The susceptible boundaries are assumed to fail at a small crack opening displacement when encountered by the propagating crack. The strain at failure, ε_f, is the property that determines the mechanical behaviour of these susceptible boundaries. The resistant boundaries are assumed to be elastic-plastic with linear isotropic hardening described by the following parameters: yield stress $\sigma_y = 200$ MPa at yield strain $\varepsilon_y = 10^{-3}$; ultimate strength $\sigma_u = 400$ MPa at ultimate strain $\varepsilon_u = 0.1$. The constitutive boundary properties are illustrated in Fig. 2. In summary, all boundaries have identical elastic properties, but the susceptible ones fail completely if the prescribed critical strain is reached, while the resistant ones yield. For a given fraction of susceptible boundaries a random distribution of these boundaries are applied to the structural elements.

The load is symmetric and applied via prescribed displacements: $u_1 = -0.025D$ along the boundary $\{X_1 = -50D,\ 0 \le X_2 \le 50D\}$ and $u_1 = 0.025D$ along the boundary $\{X_1 = 50D,\ 0 \le X_2 \le 50D\}$. Displacements $u_2 = 0$ are prescribed along the boundary $\{-50D \le X_1 \le 50D,\ X_2 = 50D\}$, while zero stresses are prescribed for all other boundary conditions. This introduces a homogeneous strain in the assembly $\varepsilon_\infty = 5\text{x}10^{-4}$, equivalent to a homogeneous stress $\sigma_\infty \approx 0.5\sigma_y$.

Two types of crack advance behaviour have been considered: Type A, where no bridging by ductile ligaments (resistant boundaries) is allowed and the crack arrests when it cannot rupture any boundary adjacent to the crack surface; and Type B, where the crack is effectively allowed to by-pass resistant boundaries and propagate further if a susceptible boundary from the first subsurface layer adjacent to the arrested crack tip could be ruptured, i.e. cracks which would arrest in Type A may be able to propagate further in type B, leaving a ductile ligament behind the crack tip. In Type B, a reduced rupture strain, ε_r, is used for the failure criterion of subsurface boundaries. This accounts for the propagation of the crack in three dimensions around the resistant boundary, which would locally raise the strain at the subsurface boundary in the 2D model. In order to define the reduced rupture strain used for Type B propagation, it is important to estimate the strain at the midpoints (which are also the integration points) of the beam elements in the vicinity of the crack tip. This can be approximately done by imagining a straight crack in a homogeneous body under plane strain conditions with the same X_2-extension as the intergranular crack. Fig. 3 illustrates the argument for the two possible cases of the imaginary crack position, denoted by 1 and 2 respectively (see also Fig. 1). For this purpose, it is useful to express the tangential strains in polar coordinates (e.g. [20]):

$$\varepsilon_t = \frac{1-\nu^2}{E}\frac{K_I}{\sqrt{2\pi r}}\left[\cos^3\frac{\theta}{2} - \frac{\nu}{1-\nu}\cos\frac{\theta}{2}\left(1+\sin^2\frac{\theta}{2}\right)\right] \qquad (2)$$

where K_I is the current stress intensity, r and θ are the polar coordinates of the point in question with respect to a polar coordinate system centred at the tip and having a zero axis parallel to crack propagation direction.

For the closest boundary in case 1, the tensile strain in the corresponding beam element is approximately equal to ε_{11}, which could be calculated from Eqn (2) with $r = D / 4$ and $\theta = 0$. For the closest boundaries in case 2, the tensile strain in the corresponding beam element is approximately equal to ε_{21}, calculated from Eqn (2) for $r = D / 8$ and $\theta = \pi / 3$. From (1) and (2), for example, the initially introduced crack in Fig. 1 causes a tensile strain in the two adjacent inclined beam elements approximately equal to $0.5\varepsilon_\infty$. This is a conservative estimate, but offers an indication of the susceptible boundaries' failure strain. The use of a much larger ε_f would make crack advance impossible.

In a similar but more elaborate way, the strains in the subsurface boundaries, ε_{12}, and ε_{22}, could be calculated. In Type B crack advance, if the boundaries adjacent to the tip are resistant, it is assumed that the crack bypasses them, mimicking 3D behaviour, and arrives at point 1' or 2' in Fig. 3. Then the real strain in the subsurface boundaries for case 1 would be approximately ε_{21}, while the real strain in the subsurface boundaries for case 2 would be approximately ε_{11} (neglecting the small change in stress intensity due to crack advance). In order to simulate this situation in the 2D settings, the failure strain of the subsurface boundaries should be reduced by a factor $\alpha_1 = \varepsilon_{12} / \varepsilon_{21} \approx 1.0$ for position 1 and by a factor $\alpha_2 = \varepsilon_{22} / \varepsilon_{11} \approx 0.77$ for position 2, respectively. To simplify the calculations a reduction factor $\alpha = (\alpha_1 + \alpha_2) / 2$ is chosen for all subsurface boundaries, so that $\varepsilon_r = \alpha\, \varepsilon_f$.

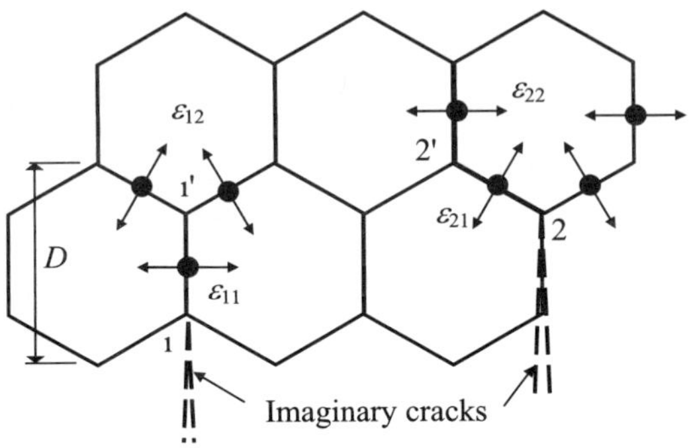

Figure 3: Estimation of grain boundary strains.

3 Crack advance simulations

The problem of crack propagation is split into evolution and equilibrium parts and these are handled by two separate software components. The first component, an in-house made program, is used for preparation of initial geometry, boundary conditions and a randomised distribution of the structural members' properties. The boundary value problem thus formulated is solved by the commercial finite element program ABAQUS [21]. The solution delivered by ABAQUS is used by the in-house program, which traverses all elements with nodes belonging to the crack surface to localise the most strained element. The decision on whether the crack will propagate is based on the following rules:

(1) If the most strained element is susceptible and the tensile strain $\varepsilon > \varepsilon_f$, then the element is removed from the structure and the simulation continues.

(2) If the most strained element is resistant and yields, then:

- (2a) For series A the crack is assumed arrested and the simulation terminates.
- (2b) For series B the elements in its immediate vicinity under the crack surface are considered; if the most strained of them has a tensile strain $\varepsilon > \varepsilon_r$, then this element is removed from the structure and the simulation continues. If there is no such element the crack is arrested and the simulation terminates.

In case of crack continuation, at most one structural member is allowed to fail and is then removed from the structure. This changed geometry constitutes a new boundary value problem, which is formulated by the in-house program and solved by ABAQUS. The process is repeated either until crack arrest or until the crack X_2-extension reaches a prescribed limit of half the assembly, $L = 25D$. It is assumed that a crack reaching the limit L fractures the assembly.

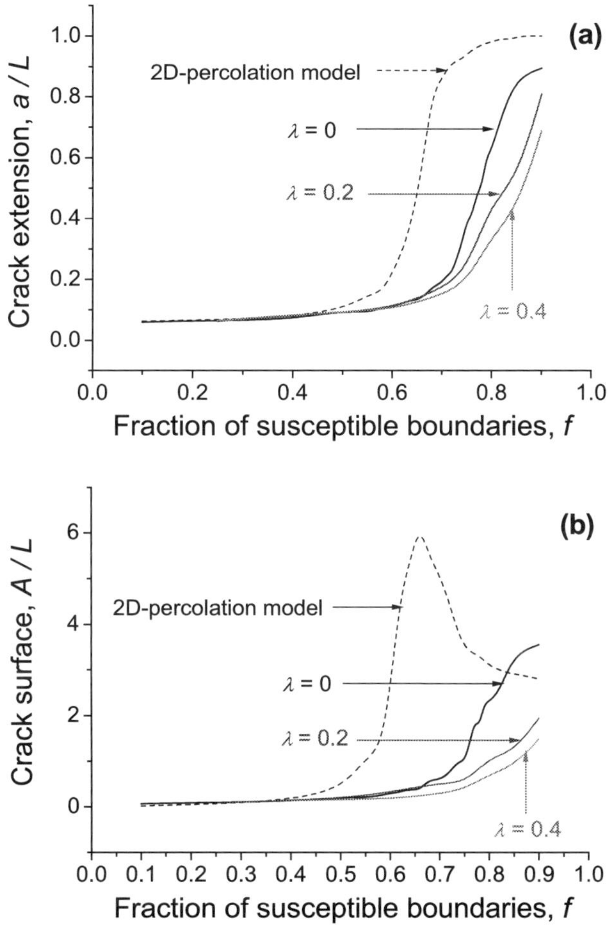

Figure 4: Expected crack extension (a) and expected crack area (b) as a function of susceptible boundaries fraction for Type A advance.

4 Results and discussion

4.1 2D simulations: type A

For 2D simulations without bridging, attention was focussed on the differences between strain-directed crack propagation in comparison with a pure percolation model. Simulations were performed for a series of susceptible boundary fractions, f = *(Number of susceptible boundaries / Total number of boundaries)*, ranging from 0.1 to 0.9. For each f, a Monte Carlo type of solution was utilised, with 30 different random distributions of susceptible boundaries. Each of these determined a separate crack evolution problem which was solved as described in the previous section. For comparison, pure percolation simulations for the same fractions and distributions were also performed. The parameters monitored for each finite element evolution problem or crack percolation problem were the final crack extension (the crack projection on X_2-axis) and the total crack area (given by total number of ruptured boundaries, i.e. including all crack branches). For each f, the average of crack extensions and the average of crack areas from the 30 random distribution solutions was found, and is taken to represent the expected crack extension, a, and crack area, A, for this particular fraction of susceptible boundaries. Results for the crack extensions and crack areas are presented in Fig. 4a and Fig. 4b, respectively, for the pure percolation model and for the finite element model with three choices of susceptible boundary rupture strain, given by $\lambda = \varepsilon_f / \varepsilon_\infty = 0;\ 0.2;\ 0.4$. The results are normalised with respect to the limiting crack length, L. Note that the simulations with $\lambda = 0$ correspond to the early percolation models based on GBCD and favourable orientation of grain boundary with respect to applied stress [12,13]. It is evident for the case $\lambda = 0$ that the role of mechanical strain is to direct the crack growth by decreasing crack branching. There is also an increase in the fraction of susceptible boundaries needed to develop a crack of given extension compared to the pure percolation model. This case is effectively a limiting solution for infinite applied stress. For finite values of remotely applied stress, any increase of the strength of susceptible boundaries increases the fracture resistance of the assembly, by increasing the fraction needed for given crack extension at a given applied stress.

4.2 Quasi 3D simulations: type B

The model with bridging allows more realistic simulations of crack propagation. It is expected that the model should show intermediate behaviour between pure 2D and 3D percolation behaviour. Therefore, full 3D percolation simulations using assemblies of tetrakaidecahedrons were also performed for comparison. A tetrakaidecahedron is a body bounded by six squares and eight regular hexagons, having cubic and octahedral symmetry. As this structure cannot be directly represented in 2D, the 3D percolation results may be used as a qualitative illustration of the expected behaviour of a system in which fully three-dimensional bridging is possible. As in the previous 2D model, finite element simulations were performed for f ranging from 0.1 to 0.9, and for each fraction 30 different random distributions of susceptible boundaries were considered. The results of all simulations are summarised in Fig. 5, where the same notations and scaling conventions as in Fig. 4 are used. The results show that the bridging indeed gives behaviour closer to the 3D percolation, with a major controlling parameter being the failure strain of the susceptible boundaries, ε_f, for given applied strain, ε_∞. Such a parametric study could be used to find an average value for ε_f from experimental results for crack length distributions, potentially turning the 2D-bridging model into predictive tool. The application of a true 3D tetrakaidecahedral model along the same lines as in this work is currently limited by the computational resources available. A small-scale model could be useful for tuning the 2D-bridging model with information on crack bridging behaviour. The 2D model could then be used on larger grain assemblies to investigate crack behaviour in stress gradients, for example.

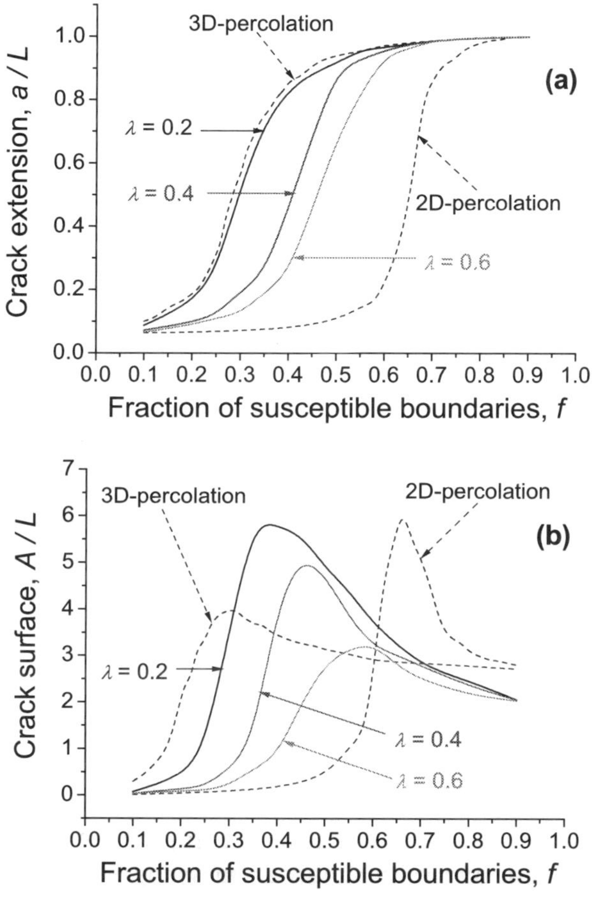

Figure 5: Expected crack extension (a) and expected crack area (b) as a function of susceptible boundaries fraction in Type B advance.

5 Conclusions

The proposed discrete structural model has the potential to simulate intergranular crack propagation in a realistic manner by including the phenomenon of crack bridging by ductile ligaments. It is applicable to sufficiently large grain aggregates in two dimensions to represent an entire test specimen. It accounts for the effects of external load magnitude and failure properties of susceptible boundaries. These effects are demonstrated in Fig. 4 and Fig. 5 in comparison to the results of previous percolation models. Parametric studies of the type presented in the work related to experimental observations of intergranular stress corrosion crack extensions and distributions are expected to tune the model parameters.

Acknowledgment

The authors gratefully acknowledge the support of this project by Rolls Royce plc.

References

[1] Watanabe T, *Res Mechanica* 11 (1): 47-84, 1984.
[2] Lim LC, Watanabe T, *Acta Metall. Mater.* 38 (12): 2507-2516, 1990.
[3] Watanabe T, *Mater. Sci. Engng.* A176 (1-2): 39-49, 1994.
[4] Watanabe T, Tsurekawa S, *Acta Mater.* 47 (15-16): 4171-4185, 1999.
[5] Lin P, Palumbo G, Erb U, Aust KT, *Scripta Metall. Mater.* 33 (9): 1387-1392, 1995.
[6] Lehockey EM, Palumbo G, Lin P, Brennenstuhl AM, *Scripta Mater.* 36 (10): 1211-1218, 1997.
[7] Gertsman VY, Bruemmer SM, *Acta Mater.* 49 (9): 1589-1598, 2001.
[8] Garbacz A, Grabski MW, *Acta Metall. Mater.* 41 (2): 469-473, 1993.
[9] Gertsman VY, Tangri K, Valiev RZ, *Acta Metall. Mater.* 42 (6): 1785-1804, 1994.
[10] Randle V, *Acta Metall. Mater.* 42 (6): 1769-1784, 1994.
[11] Davies P, Randle V, Watkins G, Davies H, *J. Mater. Sci.* 37: 4203-4209, 2002.
[12] Palumbo G, King PJ, Aust KT, Erb U, Lichtenberger PC, *Scripta Metall. Mater.* 25 (8): 1775-1780, 1991.
[13] Aust KT, Erb U, Palumbo G, *Mater. Sci. Engng.* A176 (1-2): 329-334, 1994.
[14] Gertsman VY, Janecek M, Tangri K, *Acta Mater.* 44 (7): 2869-2882, 1996.
[15] Gertsman VY, Tangri K, *Acta Mater.* 45 (10): 4107-4116, 1997.
[16] Lehockey EM, Brennenstuhl AM, Thompson I, *Corr. Sci.* 46 (10): 2383-2404, 2004.
[17] Engelberg DL, Marrow TJ, Newman RC, Babout L, In: *Proc. 2nd Int. Conf. Environment-induced Cracking of Metals*, 2004 (in press).
[18] Thomas LE, Bruemmer SM, *Corrosion* 56: 572-587, 2000.
[19] Marrow TJ, Babout L, Connolly BJ, Engelberg D, Johnson G, Buffiere J-Y, Withers PJ, Newman RC, In: *Proc. 2nd Int. Conf. Environment-induced Cracking of Metals*, 2004 (in press).
[20] Tada, H., Paris, P.C. & Irwin, G.R., *The stress analysis of cracks handbook*, 3d Ed. ASME Pres: New York, 2000.
[21] ABAQUS User's Manual, Version 6.4, Abaqus Inc., 2004.

Stress-corrosion of cold drawn prestressing steels

J. Toribio
Department of Materials Engineering, University of Salamanca, Spain

Abstract

This paper evaluates the anisotropic stress-corrosion behaviour of high-strength cold-drawn prestressing steel wires. To this end, two eutectoid steels in the form of a hot rolled bar and cold drawn wire were tested. While a tensile crack in the hot rolled bar always propagates in mode I, in the cold drawn wire an initially mode I crack deviates significantly from its normal mode I growth plane and approaches the wire axis or cold drawing direction, thus producing a mixed mode propagation. In hydrogen-assisted cracking the deviation happens just after the fatigue pre-crack, whereas in localized anodic dissolution the material is able to undergo mode I cracking before the deflection takes place. An explanation of such behaviour can be found in the pearlitic microstructure of the steels. This microstructural arrangement is randomly oriented in the case of the hot rolled bar and markedly oriented in the wire axis direction in the case of the cold drawn wire. Thus both materials behave as composites at the microstructural level and their plated structure (oriented or not) would explain the different behaviour in a corrosive environment.
Keywords: stress-corrosion cracking, pearlitic steel, steelmaking, manufacturing, cold drawing, anisotropic fracture behaviour, fracture micromechanisms.

1 Introduction

Stress corrosion cracking (SCC) of high-strength prestressing steel is a problem of major technological concern since it can increase the risk of failure of concrete structures. This paper analyses a wide set of SCC laboratory experiments of pearlitic steel wires before and after cold drawing, i.e., the previous (base) hot rolled material and the fully drawn prestressing steel wire (final commercial product used in prestressed concrete, a material of the highest interest in structural engineering). A discussion is presented to rationalise the anisotropic SCC behaviour of cold drawn prestressing steel on the basis of its oriented pearlitic microstructure, to formulate micromechanical models of macroscopic SCC behaviour and to provide the engineer with design tools in damage tolerance and structural integrity analyses of prestressed concrete structures.

2 Materials and effect of cold drawing

A high strength eutectoid steel supplied from commercial stock by EMESA was used in this work. It is the steel used in the production of cold drawn wire for prestressed concrete. The chemical composition is given in Table 1. This steel was tested in two conditions: firstly, as hot rolled patented cylindrical bars of 12 mm diameter, and secondly, as a commercial 7 mm diameter cold drawn prestressing wire obtained from the bar.

Table 1: Chemical composition (wt %) of the steel.

Mo	C	Mn	Si	P	S	Cr	Ni
0.74	0.70	0.20	0.016	0.023	0.01	0.01	0.001

The mechanical properties of both the bar and the wire are presented in Table 2. The fracture toughness K_{IC} was determined using cylindrical pre-cracked specimens obtained from the bar and the wire—for which the plane strain condition is achieved at the inner points of the crack—together with an expression for the maximum stress intensity factor at the deepest point of the crack (assumed semi-elliptical) calculated by using the Finite Element Method (FEM) combined with a Virtual Crack Extension technique.

Table 2: Mechanical properties of the bar and the wire.

Steel	Young's Modulus (GPa)	Yield Strength (MPa)	U.T.S. (MPa)	Elongation under UTS (%)	Fracture toughness (MPa $m^{1/2}$)
1. Hot rolled bar	195	725	1300	8.0	53
2. Cold drawn wire	190	1500	1830	5.8	84

The microstructure of both steels consists of fine pearlite with an interlamellar spacing of 0.1 mm. Fig. 1 shows the microstructure of both the hot rolled patented bar and the cold drawn wire in transverse and longitudinal cross sections. While the hot rolled bar has a randomly-oriented microstructure in both transverse and longitudinal sections (Figs. 1a and 1c), the cold drawn wire presents a randomly-oriented appearance in the transverse cross section (Fig. 1b), but a marked orientation in the longitudinal cross section (Fig. 1d), which implies an effect of manufacturing on the resulting microstructure. Thus the cold drawn wire presents features consisting mainly of alternate lamellae of ferrite and cementite aligned parallel or quasi-parallel to the wire axis or cold drawing direction. The consequence will be a highly anisotropic behaviour.

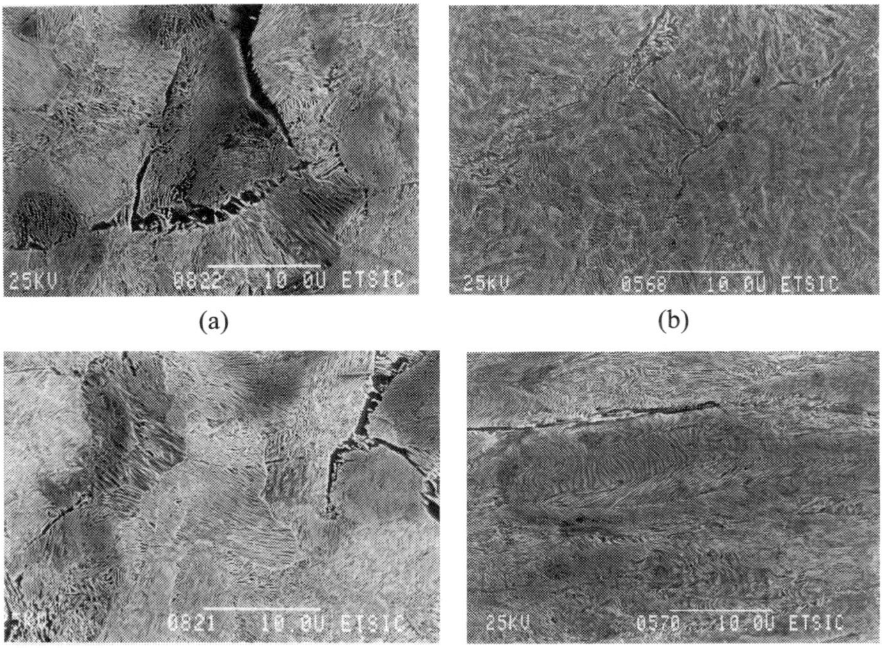

Figure 1: Microstructure of both the hot rolled and the cold drawn steels in transverse and longitudinal metallographic sections: (a) hot rolled-transverse, (b) cold drawn-transverse, (c) hot rolled-longitudinal, (d) cold drawn-longitudinal.

3 Stress corrosion tests

To evaluate the stress-corrosion behaviour of the steels in aggressive media, slow strain rate tests (SSRT) were performed on transversely pre-cracked rods immersed in aqueous environment under electrochemical control. Precracking of the samples was carried out by axial fatigue in air environment, using different fatigue loads during the last step. The maximum stress intensity factor (*K*)-levels in fatigue were K_{max} = 0.28 K_{IC}, 0.45 K_{IC}, 0.60 K_{IC} and 0.80 K_{IC}, where K_{IC} is the fracture toughness of the material in air.

3.1 Engineering approach: fracture load

The macroscopic effects of the environment on fracture were quantified through the ratio of the failure load in the solution (critical value F_C) to the failure load in air (reference value F_0), as depicted in Fig. 2. All results showed the well known anodic and cathodic regimes of environment-sensitive cracking: for higher potentials (E = –400 mV SCE) the *anodic regime,* associated with localized anodic dissolution (LAD); for lower potentials (E = –1200 mV SCE) the *cathodic regime*, associated with hydrogen assisted cracking (HAC). Since the results do not substantially depend on pH, only the average results for the three pH values used in the tests are shown in the plot of Fig. 2.

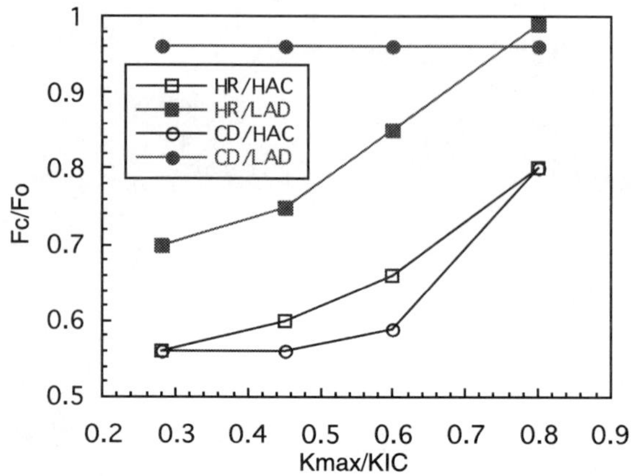

Figure 2: Macroscopic results of the slow strain rate tests, quantified through the ratio of the failure load in the solution (critical value F_C) to the failure load in air (reference value F_0), for both materials (HR: hot rolled, CD: cold drawn) and the two environmental conditions (HAC: hydrogen assisted cracking, LAD: localized anodic dissolution). The plot shows the average results in the tests.

An important K_{max}-effect is observed due to compressive residual stresses in the vicinity of the crack tip during fatigue pre-cracking of the samples. This phenomenon has been discussed in previous works [1], the main conclusion being that high values of K_{max} produce strong compressive residual stresses in the vicinity of the crack tip, thus delaying the hydrogen entry (in HAC) or the metal dissolution (in LAD).

Both the hot rolled bar and the cold drawn wire are more susceptible to HAC than to LAD (in engineering macroscopic terms). The cold drawing process is beneficial against LAD phenomena, since it clearly increases the fracture load in the anodic regime. However, cold drawing is damaging against HAC processes, since it lowers the fracture load in a hydrogen environment (cathodic regime) for the whole range of K_{max}-values.

3.2 Physical approach: microscopic fracture modes

Fig. 3 gives the microscopic fracture modes associated with the different materials and environmental conditions. In HAC conditions (cathodic regime) the hot-rolled bar fails in mode I associated with the so called tearing topography surface (TTS) fracture mode [2,3] followed by cleavage-like propagation, whereas the cold drawn wire exhibits a shear topography with some evidence of isolated cleavage facets, and the crack approaches the axis direction producing a mixed mode stress state (longitudinal splitting or delamination). There are two embryos of fracture located symmetrically in relation to the initial crack plane (at an angle of about 80º), but only one of them becomes the final fracture path, and thus the initial *crack branching* progresses along only one of the branches, probably for statistical reasons, which makes it the fracture path of lower fracture resistance or that with the higher concentration of hydrogen.

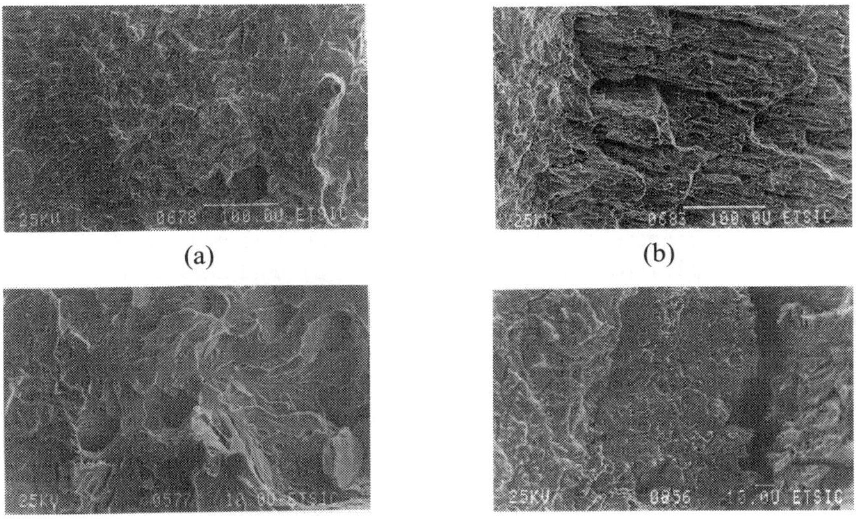

(a) (b) (c) (d)

Figure 3: Microscopic modes of fracture for an intermediate K_{max}-level of 0.45 K_{IC}: (a) hot rolled bar under HAC conditions showing TTS, (b) cold drawn wire under HAC conditions showing shear topography and isolated cleavage facets, (c) hot rolled bar under LAD conditions showing cleavage, (d) cold drawn wire under LAD conditions showing mode I dissolution and crack deflection).

Under LAD conditions (anodic regime) the hot-rolled bar fails in mode I by cleavage-like topography whereas the cold drawn wire exhibits a short mode I crack growth path (50 μm in depth), and a posterior *crack deflection* with a deviation angle of about 80º from the initial crack path in mode I. Thus the mixed mode crack growth associated with longitudinal splitting or delamination also appears in anodic (pure stress corrosion cracking) conditions.

3.3 Fracture mechanics approach: critical stress intensity factor

Important differences arise when a crack deviation from its initial plane appears and a mixed mode propagation takes place in the form of crack branching (in the case of HAC) or crack deflection (in the case of LAD), as shown in Fig. 4. As a matter of fact, the stress corrosion resistance of the cold drawn steel is strongly dependent on the capacity of it to undergo mode I cracking across the more resistant microstructural paths, since when the crack deviates from its original direction it follows a path of minimum resistance to stress corrosion cracking associated with the very oriented microstructure of the cold drawn steel.

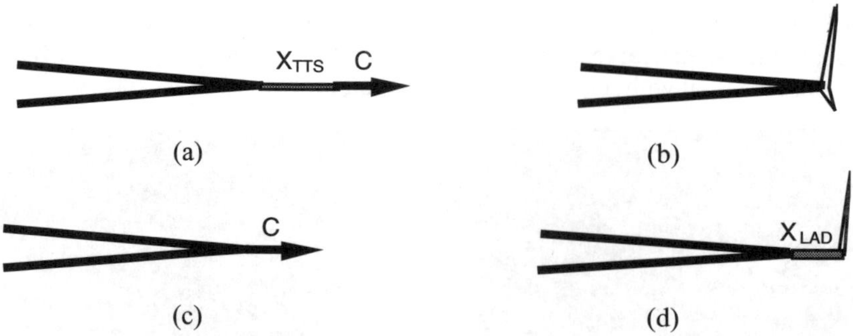

Figure 4: Schema showing the environment-sensitive cracking modes: (a) hot rolled bar in HAC conditions, (b) cold drawn wire in HAC conditions, (c) hot rolled bar in LAD conditions, (d) cold drawn wire in LAD conditions.

Table 3: Summary of results of the slow strain rate tests ($K_{max} = 0.28\ K_{IC}$). The subindex *c* indicates critical (failure) instant in the tests in aggressive solution and the subindex *0* refers to the pre-crack length before the tests (a_0) or to the reference fracture value in air (F_0, K_0).

Crack path	Env.	Mater.	Critical crack	Critical load	Critical SIF
HAC	HR	$a_C \approx a_0 + x_{TTS}$	$F_C/F_0 \approx 0.56$	K_C/K_0 – 0.56	Mode I
	CD	$a_C \approx a_0$	$F_C/F_0 \approx 0.56$	$K_C/K_0 \approx 0.56$	Mixed mode* *[Branching]*
LAD	HR	$a_C \approx a_0$	$F_C/F_0 \approx 0.70$	$K_C/K_0 \approx 0.70$	Mode I
	CD	$a_C = a_0 + x_{LAD}$	$F_C/F_0 \approx 0.96$	K_C/K_0	Mixed mode* *[Mode I+Deflection]*

HAC: hydrogen assisted cracking HR: hot rolled *: angle of 70-90º
LAD: localized anodic dissolution CD: cold drawn
TTS: tearing topography surface a: crack length
SIF: stress intensity factor K F: applied load

The environmental effects on fracture of the considered materials can be analyzed in Table 3 which summarizes the results of the slow strain rate tests for the slightest fatigue pre-cracking regime ($K_{max} = 0.28\ K_{IC}$), i.e., for that which produces the minimum previous mechanical pre-damage before the stress corrosion tests and thus allows the maximum degree of environmental damage.

In the case of HAC (cf. Fig. 4 and Table 3) the hot rolled bar exhibits better behaviour than the cold drawn wire since, although the fracture load is the same when compared with its respective value in air, the ratio of stress intensity values in aggressive medium (hydrogen) to the reference value in air is higher in the hot rolled material as a consequence of the subcritical crack growth by TTS along the original crack growth direction in mode I.

For LAD (cf. Fig. 4 and Table 3) the situation is the opposite, and in this case the effect of cold drawing on stress corrosion (LAD) performance of the steels is clearly beneficial because of the fact that the subcritical crack growth in mode I (x_{LAD}) extends the time to failure and therefore raises the fracture load before the crack deflection (and the subsequent mixed mode propagation associated with this deviation from the initial cracking path) takes place during the environment sensitive tests.

3.4 Effect of cold drawing on SCC behaviour

In HAC, both the hot rolled and the cold drawn materials suffer a marked reduction of fracture load due to hydrogen degradation, and a subsequent reduction of critical stress intensity factor K_{HAC} ("apparent toughness") which is $K_{HAC}/K_0 \approx 0.56$ in the cold drawn wire and $K_{HAC}/K_0 > 0.56$ in the hot rolled bar as a consequence of the subcritical mode I cracking by TTS (cf. Table 3).

In the case of LAD the situation is not so clear. The hot rolled bar exhibits a marked reduction of "apparent toughness" in relation to the actual toughness in air ($K_{LAD}/K_0 \approx 0.70$) in spite of the cleavage fracture micromechanism operating in both cases. As a matter of fact, although the cleavage topography is predominant in both situations, for the fracture in air (K_0) it is mechanical cleavage, whereas for the fracture in corrosive medium (K_{LAD}) it could be the so-called *film-induced cleavage* [4], that is, a brittle film which injects a crack into the substrate, thus producing cleavage fracture.

With regard to the critical stress intensity factor for LAD in the cold drawn wire, it could be even higher than the corresponding value in air ($K_{LAD}/K_0 > 0.96$). It can be explained by the blunting of the crack tip caused by the anodic dissolution that produces an apparently higher critical K-value. This crack-tip blunting is enhanced by the oriented microstructure of the drawn material (specially at the ferrite/cementite interface) and it can block the film-induced cleavage mechanism which otherwise would be operative (e.g. in the case of the hot rolled material).

The fundamental idea summarizing the results of the cold drawn steel is that such a material is highly susceptible to HAC —even more than the hot rolled initial material— but at the same time extremely resistant to LAD (without appreciable loss of fracture load). This elevated LAD resistance is associated with a clear increment of time to failure and is achieved because of the important fact that —in spite of the clear orientation of the steel microstructure— the crack is able to propagate in mode I breaking the strongest links, blunting the crack tip and blocking the film-induced cleavage mechanism. This is the key point related to the cold drawn wire: that the crack is able to undergo mode I cracking before final fracture, thus increasing both the critical stress intensity factor and the time to failure. In further sections of the paper, a micromechanical model is proposed to explain this stress-corrosion behaviour of the cold drawn steel.

4 Micromechanics of SCC in the cold drawn material

4.1 Material anisotropy

A composite engineering approach is used to model the oriented microstructure of the cold drawn wire as a *fiber-reinforced* composite. Such a microstructural arrangement has consequences of both a mechanical and a chemical nature:

(i) *Strength anisotropy (mechanical anisotropy)*, i.e., fracture toughness K_{IC} (or similar strength parameter) as a function of the orientation angle: $K_{IC} = K_{IC}(\theta)$, where θ is the common fracture mechanics angle. In particular, the fracture toughness values in direction perpendicular and parallel to the fibers are:

$$K_{IC\perp} = K_{IC}\ (\theta=0^o) \text{ for fibers fracture (transverse fracture)} \tag{1}$$

$$K_{IC//} = K_{IC}\ (\theta=90^o) \text{ for longitudinal splitting} \tag{2}$$

the former is the fracture toughness for breaking the fibers in the model (fracture of the plates or the lamellae in the real material), whereas the latter is the fracture toughness for delamination or debonding of any microstructural elements at any scale in the steel.

It may be assumed that $K_{IC\perp}$ is clearly higher than $K_{IC//}$:

$$K_{IC\perp} >> K_{IC//} \tag{3}$$

(ii) *Chemical anisotropy*, i.e., hydrogen diffusion coefficient D as a function of the orientation angle: $D = D(\theta)$, θ having the same meaning as above. This coefficient is a key item in HAC processes in pearlitic steels where stress-assisted diffusion has been shown to be the main hydrogen transport mechanism [5]. In particular, the diffusion coefficients in direction perpendicular and parallel to the fibers are respectively:

$$D_{\perp} = D\ (\theta=0^o) \text{ diffusion in the transverse (crack) direction} \tag{4}$$

$$D_{//} = D\ (\theta=90^o) \text{ diffusion in the longitudinal direction} \tag{5}$$

The former corresponds to diffusion in the crack direction, perpendicular to the fibers (or lamellae or plates, according to the modelling), and the latter is associated with diffusion parallel to the fibers.

Considering again microstructural reasons, $D_{//}$ is clearly higher than $D_{\perp}$, i.e.

$$D_{//} >> D_{\perp} \tag{6}$$

which indicates that hydrogen tends to diffuse in the direction $\theta=90^o$, thus creating a pre-damage at the microscopic level.

4.2 Micromechanisms of HAC

Hydrogen tends to diffuse, in an isotropic material, towards the points located at $\theta=0$ and the crack grows in mode I. When the material is anisotropic, such as the cold drawn wire, although the maximum hydrostatic stress is located at $\theta=0$

(polar angle form the crack tip), there are two important reasons for the crack deviation from its initial propagation path in mode I:

(i) strength anisotropy of the cold drawn wire (lower toughness in the wire axis direction)

(ii) chemical anisotropy (higher hydrogen diffusion coefficient in the wire axis direction), so that hydrogen diffuses faster in the longitudinal direction (D relatively high) than in the transverse direction ($D_{\perp}$ relatively low).

With regard to the critical mechanism of failure in HAC, two micromechanisms could be operative, according to [6]:

(i) Hydrogen enhanced localized plasticity (HELP)

(ii) Hydrogen enhanced delamination (HEDE)

The original term in [6] was hydrogen enhanced decohesion, with the same acronym HEDE, but in this work the term delamination is more adequate.

With regard to HELP, its importance in the HAC of prestressing steel is probably low, since the lamellar structure of the steel (markedly oriented, analogous to a fiber-reinforced material or a laminate at the microscopical level) probably delays or even blocks the dislocation movement which otherwise could be operative in an isotropic material. In a lamellar structure such as that of the cold drawn wire, the cementite plates act as obstacles to the movement of dislocations, so they can block (or diminish) the dislocation multiplication

In the matter of HEDE, its importance in the HAC of prestressing steel is probably high for the same reason as explained above, i.e., the lamellar structure of the steel (markedly oriented) which produces anisotropy regarding fracture and hydrogen diffusion, so that hydrogen diffuses mainly in the direction of the plates and can weaken the bonds or interfaces between the ferrite and the cementite lamellae (which are the weakest links even before the hydrogen presence) thus contributing to the hydrogen-induced fracture by delamination or debonding between two similar microstructural units.

4.3 Micromechanisms of LAD

A question arises here as to why the crack does not change its propagation path in spite of the clearly oriented microstructure of the steel. The explanation could lie in the *local (or crack-tip) strain rate* required to promote the anodic dissolution process which is achieved only at the crack tip in the θ=0 direction (i.e., the crack plane) due to the stress-strain distribution in its vicinity.

Then, why does the crack finally change its propagation direction after a certain subcritical crack growth by LAD in mode I? Because of the presence in the heavily drawn material of extremely slender *pearlitic pseudocolonies* which are potential fracture sites for two reasons:

(i) the very high local interlamellar spacing which makes them weaker or potentially fracturable by shear cracking of pearlitic plates according to the mechanism proposed by Miller and Smith [7].

(ii) the presence of some microcracks and defects consisting of plates prefractured in the pseudocolony during the manufacturing process (cold drawing) as a consequence of the very high stresses applied on the wire.

The mechanism of LAD in cold drawn steel could be explained as follows: dissolution is produced in mode I along a distance. The crack continues in mode I along the initial plane and only deviates when it the afore- mentioned pearlitic pseudocolonies and final fracture takes place for purely mechanical reasons.

5 Conclusions

The cold drawn steel is strongly susceptible to HAC —even more than the hot rolled initial material— but at the same time extremely resistant to LAD, Thus, although the stress corrosion properties of the two steels are similar in the transversal direction, the cold drawn wire exhibits a dramatic reduction of HAC resistance in the longitudinal direction.

The reason for the afore-said macroscopic behaviour is the pearlitic microstructure of the cold drawn steel which becomes oriented in the wire axis direction as a direct consequence of cold drawing. This happens at the two basic microstructural levels: the pearlitic colony and the pearlite lamellae.

Acknowledgements

The author wishes to thank the present financial support of his research at the University of Salamanca provided by the following institutions: Spanish Ministry for Scientific and Technological Research MCYT-FEDER (Grant MAT2002-01831), FEDER-INTERREG III (Grant RTCT-B-Z/SP2.P18), Junta de Castilla y León (JCYL; Grant SA078/04) and Spanish Foundation "Memoria de D. Samuel Solórzano Barruso".

References

[1] Toribio, J., Lancha, A.M. & Elices, M., Factors influencing stress corrosion cracking of high strength pearlitic steels, *Corrosion Science*, **35**, pp. 521-530, 1993.

[2] Thompson, A.W. & Chesnutt, J.C., Identification of a fracture mode: the tearing topography surface, *Metallurgical Transactions*, **10A**, pp. 1193-1196, 1979.

[3] Costa, J.E. & Thompson, A.W., Hydrogen cracking in nominally pearlitic 1045 steel, *Metallurgical Transactions*, **13A**, pp. 1315-1318, 1982.

[4] Saito, M., Smith, G.S. & Newman, R.C., Testing the film-induced cleavage model of stress-corrosion cracking, *Corrosion Science*, **35**, pp. 411-417, 1993.

[5] Toribio, J., Hydrogen-plasticity interactions in pearlitic steel: a fractographic and numerical study, *Materials Science and Engineering*, **A219**, pp. 180-191, 1996.

[6] Van Leeuwen, H.P, The kinetics of hydrogen embrittlement: a quantitative diffusion model, *Engineering Fracture Mechanics*, **6**, pp. 141-161, 1974.

[7] Miller, L.E. & Smith G.C., Tensile fractures in carbon steels, *Journal of the Iron and Steel Institute*, **208**, pp. 998-1005, 1970.

The effect of chromate on stress corrosion cracking of type 316 austenitic stainless steel in hydrochloric acid solution and the transgranular SCC mechanism

R. Nishimura & Y. Maeda
Department of Materials Science, Osaka Prefecture University, Japan

Abstract

The stress corrosion cracking (SCC) of a commercial austenitic stainless steel type 316 was investigated as a function of applied stress, chromate concentration and test temperature in 0.82 kmol/m^3 HCl solution by using a constant load method. The three parameters (l_{ss}; steady state elongation rate, t_{ss}; transition time, t_f; time to failure) obtained from the corrosion elongation curve were divided into three regions against applied stress, which were dominated by stress, SCC or corrosion. In the SCC-dominated region, log l_{ss} was a linear function of log t_f regardless of chromate concentration and test temperature, showing that l_{ss} became a relevant parameter for prediction of t_f. A critical chromate concentration and a critical test temperature, above or below which little SCC takes place, were estimated. The TGSCC mechanism was qualitatively proposed to explain the results obtained.
Keywords: type 316, chromate, stress corrosion cracking, steady state elongation rate, hydrochloric acid, corrosion elongation curve, TGSCC mechanism.

1 Introduction

It is well known that chromate is a very good inhibitor as passivator for general corrosion and localized corrosion. However, chromate is too toxic for human beings to be used in public or in service at present. In this study, we used chromate for estimating a corrosion behavior at crack tips for type 316 specimens under a constant applied stress condition, that is, acceleration of film formation or inhibition of dissolution. The present study was conducted to investigate the SCC behavior of type 316 specimens in 0.82 kmol/m^3 HCl solution with various chromate concentrations to compare with that of type 304 in 0.82 kmol/m^3 HCl solution with various concentrations of chromate and molybdate [1,2]. On the results obtained, (1) a parameter for the prediction of time to failure, and (2) critical values of chromate concentration and test temperature, below or above which no SCC takes place, were estimated and (3) a qualitative transgranular SCC mechanism was discussed.

2 Experimental

The specimens used were a commercial type 316 austenitic stainless steel (yield strength: 333 MPa and ultimate tensile strength: 647 MPa). The geometry for SCC experiments was as follows; the gauge length-- 25.6 mm, width-- 5 mm and thickness-- 1 mm. The chemical composition (mass %) was as follows: C 0.054, Si 0.67, Mn 1.38, P 0.030, S 0.005, Ni 11.16, Cr 17.21, Mo 2.21 and Fe balance. The specimens were solution annealed at 1373 K for 3.6 ks under an argon atmosphere and then water quenched. The solution annealed specimens were polished to 1000 grit emery paper, degreased with acetone in an ultrasonic cleaner and washed with distilled water. The test solutions used were 0.82 kmol/m^3 HCl solutions with a chromate concentration of 0.01 kmol/m^3 to 1.0 kmol/m^3 and were prepared from distilled water and guaranteed grade reagents. The test temperature was used in the range of 333 K to 373 K. All experiments were carried out under an open circuit condition.

A lever- type constant load apparatus (lever ratio 1:10) to which three specimens can be separately and simultaneously attached was used with a cooling system on the top to avoid evaporation of the solution during experiments. The specimens were insulated from rod and grip by surface oxidized zirconium tube. A change in elongation of the specimens under the constant load condition was measured by an inductive linear transducer with an accuracy of±0.01 mm.

3 Results

3.1 Corrosion elongation curve - determination of three parameters (t_{ss}, t_f and l_{ss})

Fig. 1 shows a representative example of the corrosion elongation curves for type 316 specimens at a constant applied stress of 395 MPa in 0.82 kmol/m^3 HCl solution with and without chromate at a test temperature of 353 K. The corrosion elongation curves, except that with a chromate concentration of 0.075 mol/m^3, consisted of three regions up to failure with an initial sudden rise of elongation; primary, secondary and tertiary regions[1- 5]. From these curves, three parameters can be obtained; (1) the steady state elongation rate in the secondary region, l_{ss}, (2) the transition time between the secondary and tertiary regions, t_{ss}, and (3) the time to failure, t_f.. If there is no failure within a laboratory time scale (< 10^7 s) such as the result at a chromate concentration of 0.075 mol/m^3 (as shown in fig.1), the tertiary region does not appear in the corrosion elongation curve and as a result only l_{ss} can be obtained and the order of the magnitude of l_{ss} becomes 10^{-10} m/s and less[1- 5].

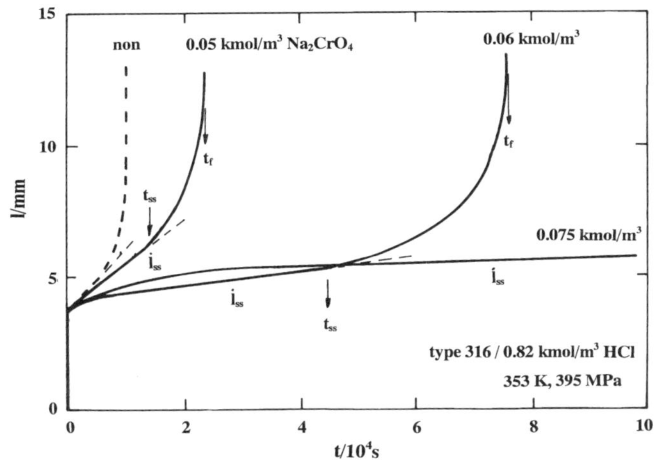

Figure 1: Corrosion elongation curves in 0.82 kmol/m^3 HCl solution with various chromate concentrations at a test temperature of 353 K under a constant applied stress of 395 MPa.

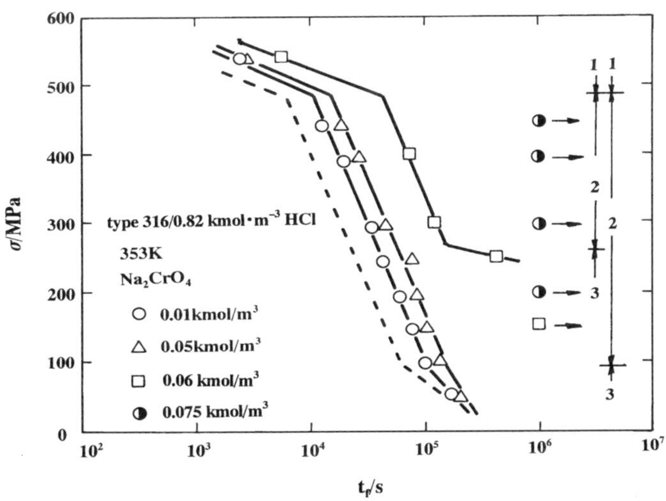

Figure 2: The relationships between applied stress and time to failure in 0.82 kmol/m^3 HCl solution with various chromate concentrations, where the arrows show no failure.

3.2 Applied stress dependence of three parameters

Fig. 2 shows the applied stress dependence of t_f at various chromate concentrations in 0.82 kmol/m^3 HCl solution at 353 K, where the dashed line is the result without chromate and the arrows show no failure. The applied stress versus t_f curves except a chromate concentration of 0.075 mol/m^3 were composed of three regions (stress- dominated, SCC- dominated and corrosion-dominated regions) as well as that without chromate. At 0.06 mol/m^3, the applied stress range in the SCC- dominated region became narrow, while at 0.075 mol/m^3, there was no failure at any applied stresses in the SCC- dominated region. The applied stress dependence of l_{ss} corresponded to that of t_f in fig. 2 and the value of l_{ss} of the specimens shown by the arrows in fig. 2 was of the order of 10^{-10}m/s and less. In addition, from the relationships between applied stress and the ratio of t_{ss} to t_f,, the value of t_{ss}/t_f in the SCC- dominated region held constant with a value of 0.57 $\pm$ 0.02 independent of applied stress and chromate concentration, while the values in the stress- dominated and corrosion-dominated regions became higher than that in the SCC- dominated region.

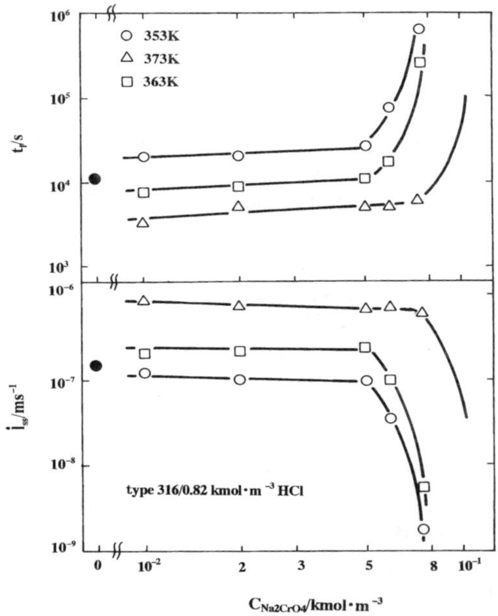

Figure 3: The relationships between t_f,/ l_{ss} and chromate concentration at various constant test temperatures under a constant applied stress of 395 MPa in 0.82 kmol/m^3 HCl solution.

3.3 Chromate concentration and test temperature dependences of three parameters

Fig. 3 shows an example of the relationships between t_f,/ l_{ss} and chromate concentration at various constant test temperatures under a constant applied stress of 395 MPa in 0.82 kmol/m^3 HCl solution, where similar results were obtained at other constant applied stresses. The slight increase in t_f or decrease in l_{ss} took place with increasing chromate concentration. At or above a certain chromate concentration, t_f or l_{ss} rapidly increased or decreased. The value of t_{ss}/t_f was constant with a value of 0.57 $\pm$ 0.02 independent of test temperature and chromate concentration. In addition, similar results were obtained for the relationships between three parameters (t_f, l_{ss} and t_{ss}/t_f) and a reciprocal of test temperature at various constant applied stresses in 0.82 kmol/m^3 HCl solution with various chromate concentrations With decreasing test temperature t_f increases and l_{ss} decreases linearly and then the rapid increase in t_f or decrease in l_{ss} took place at or below a certain test temperature. The value of t_{ss}/t_f was held constant with a value of 0.57 $\pm$ 0.02 independent of applied stress and test temperature

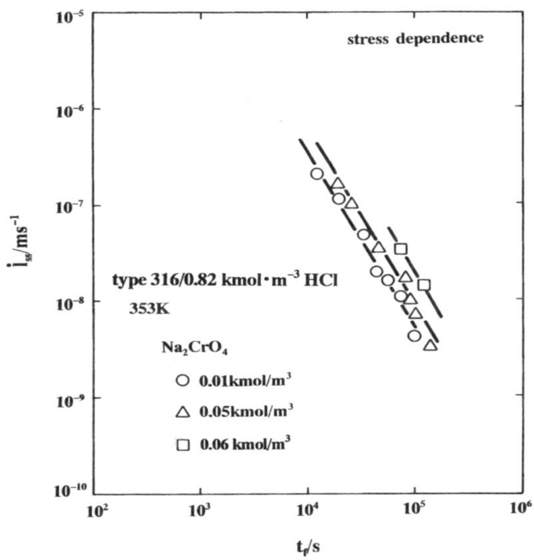

Figure 4: The relationships between l_{ss} and t_f obtained at various chromate concentrations as a function of applied stress in the SCC-dominated region.

4 Discussion

4.1 A parameter for prediction of time to failure

Fig. 4 shows the relationships between log l_{ss} and log t_f obtained as a function of applied stress in the SCC- dominated region at various constant chromate concentrations. The relationships became the good straight lines with a slope of - 2 irrespective of chromate concentration. Therefore, we obtained the following empirical equation,

$$\log l_{ss} = -2\log t_f + C_1 \quad (\sigma\text{: variable}) \tag{1}$$

where C_1 is a constant depending on chromate concentration.

The relationships between log l_{ss} and log t_f, obtained as functions of chromate concentration and test temperature under a constant applied stress condition became a good straight line with a slope of - 1 irrespective of chromate concentration and test temperature. We obtained the following empirical equation,

$$\log l_{ss} = -\log t_f + C_2 \quad (\sigma\text{: constant}) \tag{2}$$

where C_2 is a constant depending on applied stress, but independent of the environmental factors.

The linear equations in eqns (1) and (2) imply that l_{ss} becomes a relevant parameter for predicting t_f independent of applied stress, chromate concentration and test temperature, because l_{ss} can be obtained at a time within 10 to 20% of t_f from the corrosion elongation curve. On the other hand, the specimens had no SCC susceptibility within the laboratory time scale, when the value of l_{ss} became in the order of 10^{-10} m/s and less. This implies that l_{ss} also becomes a parameter for the assessment of whether SCC takes place or not.

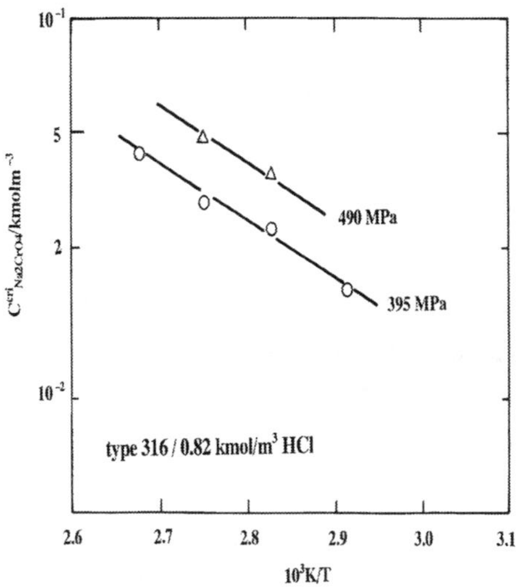

Figure 5: Critical chromate concentrations as a function of test temperature at the constant applied stresses of 395 MPa and 490 MPa.

4.2 Critical values of chromate and test temperature

From fig. 3, log t_f or log l_{ss} began to increase or decrease significantly above a certain chromate concentration, which concentration was roughly defined as a critical chromate concentration ($C^{cri}{}_{Na2CrO4}$) at each test temperature. The $C^{cri}{}_{Na2CrO4}$ increased with increasing test temperature. Fig. 5 shows the relationships between $C^{cri}{}_{Na2CrO4}$ and a reciprocal of test temperature at the constant applied stresses of 395 MPa and 490 MPa. The relationships became a good straight line with a same slope. No SCC would take place at any applied stresses in the SCC-dominated region above $C^{cri}{}_{Na2CrO4}$ at 490 Mpa, and then was expressed as,

$$\log C^{cri}{}_{Na2CrO4} = -\ 1469/T + 3.047 \quad (\sigma\text{: 490 MPa}) \tag{3}$$

A critical test temperature at each chromate concentration ($T^{cri,\ \sigma}$) was estimated as a function of applied stress as well as $C^{cri}{}_{Na2CrO4}$. The relationships between applied stress and $T^{cri,\ \sigma}$ became a good straight line and then we got T^{cri} at 490 MPa at each chromate concentration. The relationship between T^{cri} and chromate concentration became a good straight line and expressed as,

$$10^3/\ T^{cri} = -\ 0.35 \log C_{Na2CrO4} + 2.52 \quad (\sigma\text{: 490 MPa}) \tag{4}$$

4.3 A proposal of the transgranular SCC (TGSCC) mechanism

The fracture appearances were transgranular regardless of applied stress, chromate concentration and test temperature, whenever SCC took place. Therefore, the TGSCC mechanism [2] is based on a film rupture-formation event with a time interval of Δt at crack tips which are nucleated at slip steps emerging on the surface. A film formed at crack tips inhibits dislocation movement to produce pile- ups of dislocation and as a result an additional local stress (σ_{dis}) in the vicinity of the cracks is developed in addition to a local stress (σ_{app}) caused by a constant applied stress (σ). On the other hand, a dissolution current density (j_d) at crack tips accelerates dislocation movement by injection of defects into the dislocation pile- ups [6]. It is assumed that the film rupture takes place when a net local stress at crack tips (σ_{tip}) reaches a critical fracture stress (σ_F), and σ_{tip} decreases simultaneously up to σ_{app} just after film rupture,

$$\sigma_{tip} = \sigma_{app} + \sigma_{dis} \quad (\sigma_{app} \leqq \sigma_{tip} \leqq \sigma_F\ ) \tag{5}$$

As σ_{app} is presumed to keep immediately constant just after the application of load, $d\sigma_{app}/dt$ is zero; $d\sigma_{tip}/dt = d\sigma_{diis}/dt$ during film formation. Finally, we have obtained the following relationship [2],

$$d\sigma_{tip}/dt \propto (\ \tau\ -\ \tau_i\ )\ (L_{tip}/r)^{-1/2}\ j_d/zF \tag{6}$$

where τ is the shear stress at σ (σ= 2 τ), τ_i is the friction stress and L_{tip} is the length of piled- up dislocation at σ_{tip} and r is the length from the tip of the pile up which would correspond to a film thickness at crack tips.

Thus, the increasing rate of σ_{tip}, $d\sigma_{tip}/dt$, is proportional to τ (or σ) and j_d, assuming that L_{tip}/r keeps approximately constant. Under a constant applied stress (τ = constant), $d\sigma_{dis}/dt$ increases with increasing j_d. Therefore, the time, Δt, for σ_{tip} to reach σ_F decreases with the increase in $d\sigma_{dis}/dt$ followed by the increase in j_d. On the other hand, when σ is increased under a constant environmental condition (approximately the behavior of j_d is the same), τ increases correspondingly. This means that the contribution of σ_{dis} to σ_{tip} decreases; in other words, σ_{tip} reaches σ_F by the slight increase in σ_{dis}. If σ_{app} becomes equal to σ_F or more by the increase in applied stress, a mechanical fracture would take place with little contribution of σ_{dis} , which corresponds to the stress- dominated region.

On the basis of the above considerations, using eqn (6), we can explain the effect of chromate and test temperature on SCC susceptibility. It is well known that the chromate serves as a passivator which makes formation of a passive film easy. As the concentration of chromate increases, a repassivation rate after film rupture increases. Consequently, j_d becomes small and $d\sigma_{dis}/dt$ decreases correspondingly. This means that the decrease in $d\sigma_{dis}/dt$ leads to the increase in Δt, which corresponds to the decrease in SCC susceptibility. As test temperature increases, the repassivation rate decreases and hence j_d becomes large for $d\sigma_{dis}/dt$ to increase correspondingly.

5 Conclusions

(1) The relations between log l_{ss} and log t_f obtained as functions of applied stress, chromate concentration and test temperature become good straight lines. This implies that the l_{ss} becomes the parameter for prediction of time to failure t_f.

(2) The critical values of chromate concentration and test temperature were estimated as follows,

$$\log C^{cri}{}_{Na2CrO4} = -\ 1469/T + 3.047 \quad (\sigma\text{: 490 MPa,})$$

$$10^3/\ T^{cri} = -\ 0.35 \log C_{Na2CrO4} + 2.52 \quad (\sigma\text{: 490 MPa})$$

(3) The qualitative TGSCC mechanism was proposed to explain the effects of chromate and test temperature on SCC susceptibility.

References

[1] R. Nishimura and Sundjono, Corrosion, **56**(2000) 361- 370.

[2] R. Nishimura and Y. Maeda, Corros. Sci., **46** (2004)343- 360.

[3] R. Nishimura and K. Kudo, Corrosion, **45**(1989) 308- 316.

[4] R. Nishimura, Corrosion, **46**(1990) 311- 318.

[5] R. Nishimura, Corrosion, **49**(1993) 796- 801.

[6] B. Gu, W.-Y. Chu, L.- J. Qiao, C.- M. Hsiao, Corros. Sci., **36**(1994)1437-1444.

Author Index